GREAT AMERICAN HOMES

PLANTATIONS OF THE OLD SOUTH

BY HENRY WIENCEK
PHOTOGRAPHY BY PAUL ROCHELEAU

Oxmoor
House®

Great American Homes
was created and produced by
Rebus, Inc.
and published by Oxmoor House, Inc.

Rebus, Inc.
Publisher: Rodney Friedman
Editor: Charles L. Mee, Jr.
Senior Picture Editor: Mary Z. Jenkins
Picture Editor: Deborah Bull
Art Director: Ronald Gross
Managing Editor: Fredrica A. Harvey
Consulting Editor: Michael Goldman

Production: Paul Levin,
Giga Communications, Inc.

Author: Henry Wiencek, currently
editor of *The Smithsonian Guide to
Historic America,* is the author of
Mansions of the Virginia Gentry (another
of the Great American Homes
volumes). He has also written histories
of Japan, Portugal, and Mexico.

Photographer: Paul Rocheleau is a
Massachusetts-based photographer
whose work has appeared in *Antiques,
Architectural Digest,* and *Americana*
magazines.

Consultant: Arthur Edwards—born
and bred in North Carolina—has taught
architectural history at the Pratt
Institute in New York since 1972.
Previously a member of the North
Carolina state historic preservation
agency, he has developed travel
programs to study architectural history
in America and abroad.

Published by Oxmoor House, Inc.
Book Division of Southern Progress Corporation
P.O. Box 2463
Birmingham, AL 35201

Library of Congress
Cataloging in Publication Data
Wiencek, Henry.
  Plantations of the Old South.

  (Great American homes)
  Includes index.
  1. Plantations—Southern States. I. Title.
II. Series.
NA7211.W5   728.8′3′0975   '84-17081
ISBN 0-8487-0758-3

Cover: Rosedown Plantation.

CONTENTS

FOREWORD

ROSEDOWN PLANTATION

No architectural type is more quintessentially Southern than a plantation house. The antebellum home with white columns framed by lush plantings of trees and shrubs has become the staple image of gracious living in a bygone era. Nostalgia has distorted the vision, conveniently overlooking the unjust social system that made possible luxurious existence for a few; slave quarters do not figure prominently in the idealized plantation picture. But is was precisely the combination of forced labor and fertile soil that supported the plantation economy, and after 1830 affluent landowners across the Deep South aspired to express their wealth and culture through the construction of magnificent homes suited for expansive living and formal entertainment. Many chose the then-fashionable Greek Revival style because its symmetrical qualities conferred dignity and symbolically linked the new democracy of the United States with its spiritual forebear, the Athenian democracy of the fifth century B.C. Not unimportantly, the tall columns of Greek Revival architecture accommodated that desirable subtropical feature, the shaded veranda or porch, which helped to keep the interior cool while providing a protected exterior living space.

Six prominent Southern plantation houses are described and illustrated here. The homes in this volume have been selected to reflect both geographic diversity and stylistic variety. Of the group, the Hermitage, outside Nashville, Tennessee, retains a Georgian plan, later enlarged and overlaid with a Greek Revival exterior. Waverley, near Columbus, Mississippi, has an unusual octagonal central hall rising through all three floors to a multiwindowed cupola atop the roof, while Shadows-on-the-Teche, in New Iberia, Louisiana, reflects a traditional Creole house form with a veneer of classical detailing. Rattle and Snap, near Columbia, Tennessee, is Greek Revival with an almost baroque flair, and Gaineswood, in Demopolis, Alabama, celebrates its owner-builder's free interpretation of historic precedent in a characteristically American way. All six houses can be seen today, splendidly restored and furnished much as they appeared in 1860. For modern eyes, jaded by the proliferation of tasteless suburban houses with plantation pretensions, this book serves as a reminder of the refinement, innovation, and good sense embodied in the originals.

MARIAN SCOTT MOFFETT
KNOXVILLE, TENNESSEE

INTRODUCTION

The South perfected the architecture of romantic languor. There is no place more conducive to sentimental reveries than a veranda with a view of a bayou. A brilliant row of white columns, framed by twisting live oaks, is a vision of grandeur isolated in a lush landscape. By design a Southern house is a place of soothing shadow and mild breezes, a refuge from, yet also a part of, the landscape. A Northern house is a bastion against the cold and a snug shelter from the wind; the Southern house welcomes the landscape with tall windows, broad doors, and, most of all, spacious verandas. One traveler to the South wrote that on plantations "the people *'live out of doors'* ...their very houses, ever wide open, are themselves 'out of doors.'" A Southerner could tell you exactly where the evening breeze came from—down that hill, off the bayou, there—because he built the house to catch that breeze. And in a riverfront house at night, when the breeze brings into the house the coolness, the scent, and the sound of the river, the river itself comes into the house. The house loses its substance and becomes an ethereal thing.

The landscape, as much as any other element, defines the character of the Southern mansion. So rich is the landscape, so colorful, fragrant, and fertile, that it is "a sort of cosmic conspiracy against reality in favor of romance," as Wilbur Cash said in *The Mind of the South*. "The dominant mood," he continued, "the mood that lingers in the memory, is one of well-nigh drunken reverie...of such sweet and inexorable opiates as the rich odors of hot earth and pinewood and the perfume of the magnolia in bloom—of soft languor creeping through the blood and mounting

Rattle and Snap, near Columbia, Tennessee, epitomizes the grandeur of antebellum Southern architecture. Classical colonnades such as this one proclaimed the wealth and pride of the Southern planters, who produced almost ninety percent of the world's cotton in the decade before the Civil War.

surely to the brain."

Other factors have also conspired against reality in the South. The impulse to romanticize, which colors all of American history, has wrapped the plantation in an almost impenetrable cocoon of sentimental gauze. A haze of unreality had settled over the plantation even before the Civil War. Both Northerners and Southerners, disgusted at the rapacity of merchants and speculators, saw in the Southern planter the embodiment of the old agrarian virtues, and in the plantation the alternative to the dehumanization of the factory. A corollary to this view was that the Southern slave, cared for by a benevolent master, was better off than the wage slave exploited by a Yankee capitalist in the dim and noisy Northern factory—a line of reasoning that Southerners stressed in the debates that preceded the war. The masters clung to this myth of the plantation and to the institution of slavery when the force of history was running against them. No place was more suited for fleeing reality than a plantation.

"We shall be the richest people beneath the bend of the rainbow!" exulted one Southerner in 1840, when planters were riding the crest of the cotton tidal wave. The cotton boom began about 1830, but its origin was in 1793, when Eli Whitney

invented the cotton gin (the word is short for "engine") that separated cotton fibers from the seeds. Before Whitney's gin, raising cotton was a break-even proposition at best because it took a slave two years to pick the seeds from one bale of cotton. Whitney's gin could yield fifteen bales a day and made it economical to raise cotton on a large scale. The gin breathed new life into the institution

This 1871 Currier and Ives print of a mansion on the Mississippi was based on a painting of 1850. The printmakers altered the original, draping the trees in Spanish moss—a romantic feature of the Southern landscape that Northerners had come to expect.

of slavery, which had been waning in the last decades of the eighteenth century.

In the 1820s and 1830s an increasing number of settlers from Virginia, the Carolinas, and Georgia migrated to Alabama, Mississippi, Louisiana, and Tennessee. With cotton selling for higher and higher prices every year, it was easy to buy land and slaves on credit—a plantation often paid for itself in just a few years. Most of the settlers had been small-time farmers, and they established equally modest plantations. A minority of the newcomers were relatively wealthy when they arrived;

with their larger capital they prospered on a magnificent scale, seemingly overnight. "Miraculous" was the word a South Carolina lawyer used to describe "the sudden acquisition of wealth in the cotton-growing region." "Large estates," he wrote, "as if by magic, are accumulated. The fortunate proprietors then build fine houses, and surround themselves with comforts and luxuries to which they were strangers in their earlier years of toil and care."

A group of slaves pick cotton in this painting of a typical Southern plantation. The building with the smokestack housed the cotton gin, which separated the fibers from the seeds. A steamboat has docked at the landing to pick up cotton and deliver supplies.

Though most of the plantation houses of the South were built in the same period, roughly from 1820 to 1860, they were not all built in the same style. Some houses displayed Roman elements, in keeping with the Roman revival launched by Thomas Jefferson when he completed Monticello in the early decades of the nineteenth century. Some builders, particularly in Louisiana, followed traditional local designs that were enhanced by columns and classical motifs. In the 1850s Gothic Revival and Moorish houses appeared. Oddities, such as the octagonal house, had brief vogues. Many houses were like the fictional Tara, Scarlett O'Hara's home in Margaret Mitchell's *Gone With the Wind*—"a clumsy, sprawling building...built according to no architectural plan whatsoever." But in the heyday of the cotton plantation the dominant fashion was the Greek Revival, known then as the "Grecian" style.

The Greek Revival in architecture was part of a national fascination with the Greeks, both ancient and modern, that began in the 1820s. In that decade the Greeks went to war to gain their independence from the Turks, kindling memories here of America's own struggle with Britain and reminding Americans that Greece was the birthplace of democratic ideals. All over the country young and old took up the study of the ancient Greek language to read the classics of philosophy, history, drama, and poetry. Pioneers named their new towns after the cities of the ancient world. An Englishman traveling in Tennessee wrote that "the farmers of Tennessee have their country studded with such classical names as Athens, Sparta, Troy, Carthage, Memphis, and Palmyra."

To clothe oneself in the mantle of the Greeks declared a reverence for classical rationality, enlightenment, and democracy. Nicholas Biddle, the Philadelphia banker, summed up the prevailing view when he proclaimed, "The two great truths of the world [are] the Bible and Grecian architecture." Ancient architecture also carried with it the promise of permanence—if the Parthenon had endured for so many centuries as the stone embodiment of Greek ideals, then perhaps the American system would also endure. In the South this yearning for permanence took on a special meaning. The plantation system was under attack from

After a dash through the woods a hunter prepares to finish off a stag his hounds have run to ground. "Ah! it is impossible for your pale denizens of the dusty town . . . to appreciate the wild delight of a . . . chase," said one Southerner of the pleasures of hunting.

the abolitionists, and the South drew comfort from its architectural association with an ancient and enduring civilization. And did not the Greeks keep slaves as well? An architectural style that began as a progressive, libertarian, and democratic movement on all levels of society became tinged with upper-class conservatism in the South.

The Greek Revival is seen in its purest expression in public buildings: banks, churches, courthouses, and state capitols were given not only the decorations but the very forms of temples. Convening a legislature inside a Greek temple was one thing, housing a family quite another. House builders mainly used columns to express their Grecian sentiments. Often they added an entablature—a flat band over the colonnade; less often the builders topped the colonnade with a bold triangular pediment. Though the forms of the Greek Revival were antique, Americans did not see them as being academic or stuffy—quite the contrary. In 1850 A.J. Downing wrote in *The Architecture of Country Houses*, "The prevailing character of the Greek and Italian styles partakes of the gay spirit of the drawing-room and social life....those who love sunshine and the enjoyment of the present moment will prefer the classical or modern styles." As it turned out, the classical colonnade was an architectural device perfectly suited to Southern living because it created shady and breezy porches.

The Greek Revival style was championed by the country's emerging group of professional architects, but its almost universal popularity was a grass-roots phenomenon. In the first half of the nineteenth century, the professional architect remained a rare species, one that stuck mostly to the cities. New Orleans had James

Gallier and his son, James, Jr.; Robert Mills designed several government buildings in Washington, D.C.; Ithiel Town and Alexander Jackson Davis ran a successful partnership in New York; Benjamin Henry Latrobe was active in Richmond, Washington, D.C., Baltimore, Philadelphia, and New Orleans. A lesser-known professional named David Morrison designed public and private buildings in Nashville, including Andrew Jackson's Hermitage (Chapter Five). But most plantation mansions were designed and built by their owners, sometimes with the help of a contractor. A plantation master was accustomed to doing such things on his own. If he wanted a barn or an icehouse he designed it himself, perhaps taking inspiration from illustrations in a builder's stylebook. The work would be done by slaves—either his own or those hired out from a neighbor—skilled in the crafts of construction. In some cases it is difficult to accept the fact that a gentleman farmer could have been imaginative enough and skilled enough to design a structure as lovely and as complex as Rattle and Snap (Chapter Four), for example. But the most diligent research through family letters and accounts has failed to turn up even the suggestion that George Polk had professional help in designing his mansion. In all likelihood he relied on the advice of his well-traveled brother Leonidas and may have consulted a stylebook by Minard Lafever. But no direct link between Lafever's published drawings and the design of Rattle and Snap can be proved.

Not all amateurs met with total success. Jefferson Davis, later to be the president of the Confederacy, designed his Mississippi mansion, Brierfield. His second wife, Varina Howell, wrote a forthright appraisal of her husband's work: "The building was one of my husband's experiments as an architect, and he and his friend and servant, James Pemberton, built it with the help of the negroes on the plantation. The rooms were of fair size, and opened on

On foot and in carriages spectators arrive at a Kentucky racecourse in 1840. Southerners of all social levels were keen on horse racing and gambling—they would literally bet the clothes on their backs. At one race Andrew Jackson won "$1,500 in wearing apparel."

a paved brick gallery, surrounded by latticework; but some miscalculation about the windows had placed the sills about breast high. The outer doors were six feet wide...when they were opened, the side of the house seemed to be taken down." She also thought that the mammoth fireplaces Davis ordered would have been more appropriate in an English castle.

As Varina Davis's description of her house reveals, the overriding concern of the Southern builder was the climate—he had to provide ventilation. In the eighteenth-century houses of Virginia and the Carolinas, builders had taken a simple approach to this problem: they almost always included a spacious stair hall through the center of the house with doors at the front and rear to provide a strong cross breeze. The stairway swept air up to the second floor. The Hermitage has a basically eighteenth-century floor plan with just such a central hall. Not surprisingly, General Jackson was known to take his after-dinner nap on a sofa in this hall, which was probably the coolest place in the house.

Louisiana builders had a different approach to the problem of the heat—they surrounded the house with galleries, or at the very least put a wide porch on the front of the house. The encircling galleries were an idea that may have originated in French settlements in the North—perhaps as protection for walls and foundations—and gradually made its way down the Mississippi River to Louisiana. In the houses of common farmers plain square posts supported the overhang of the roof all around the house. In the homes of the wealthy these simple posts became grand, two-story columns; and a second-floor porch, cooler and more private, was added above the ground-floor veranda.

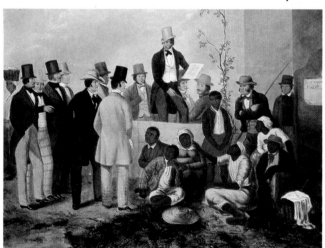

Dealers and plantation owners look over slaves offered for sale at an auction in 1852. Congress had outlawed the importation of slaves in 1808, but the domestic slave trade continued to be a lively and profitable business until the Civil War.

The floor plans devised by the Creoles—the Louisiana descendants of French and Spanish pioneers—allowed cooling without a central hallway. Shadows-on-the-Teche (Chapter One) has a typical Creole plan, even though it was built for an Anglo-Saxon, David Weeks. He placed his dining room, with doors at the front and back, at the center of the house on the ground floor. The parlor was directly above and had doorways to galleries on both sides. The bedrooms that flanked the parlor also had doors opening to the front gallery. Shadows was so breezy that Weeks's widow and her second husband decided to enclose the rear gallery and loggia.

Spacious, high-ceilinged rooms were more common in the South than in the

North, where a big room would have been chilly and dark for half of the year. Not only were large rooms appropriate in the Southern climate, but they accommodated the generous, Southern-style entertainments so beloved by the planters. George Young designed Waverley (Chapter Three) with a large rotunda for parties. General Nathan Bryan Whitfield's Gaineswood (Chapter Two) included a ballroom which he decorated with lavish plaster embellishments and two dozen pilasters. Rattle and Snap is the epitome of rural opulence: its spacious rooms accommodated scores of guests at princely entertainments.

Slaves lived in overcrowded one-room cabins, making them easy prey to disease. Rarely did a slave live to be sixty. These slave cabins in Louisiana are among the few that remain.

The plantation owners bought the finest furnishings for their mansions. Many patronized the New Orleans shops of Prudent Mallard and François Seignouret; some purchased furniture from cabinetmakers in New York and Philadelphia. In the 1820s and 1830s Empire-style furniture was fashionable in the plantation mansions; later the more exuberant Rococo Revival style was popular. Daniel and Martha Turnbull of Rosedown (Chapter Six) imported some of their furniture from Europe, as well as Carrara marble statuary for the garden. Wallpaper became popular in the early nineteenth century. Southerners had a taste for floral patterns with muted colors and occasionally for scenic paper. The Turnbulls bought wallpaper depicting the exploits of Charlemagne's knights from the French maker Joseph Dufour. Andrew Jackson purchased another Dufour creation, wallpaper that illustrated a Greek myth. The Southern taste for scenic paper sometimes descended to the garish. One visitor to a Louisiana house found herself stepping into the African jungle: "Tall trees reached to the ceiling, with gaudily striped boa-constrictors wound around their trunks; hissing snakes peered out of the jungle…monkeys swung limb from limb; orang-outangs and lots of almost naked dark-skinned natives wandered about." In another house she saw a panorama of the Ganges River, where an impoverished mother was shown tossing her infant into the jaws of a crocodile.

With its sumptuous furnishings, shady verandas, and fragrant gardens, the plantation fostered a life of ease. "It is an indolent, yet charming life, and one quits thinking and takes to dreaming," was the assessment of a young New Yorker, John Anthony Quitman, who moved to Natchez, Mississippi, in the 1820s and wrote those lines to his father during a visit to a plantation. He had fallen under the spell of the

South, a spell woven of "stately oaks," "vast, undulating sweeps of cultivated fields," and the "delightful climate." In the middle of January Quitman wrote to his undoubtedly snowbound father in upstate New York, "The peach and plum are in full bloom, and the birds sing merrily in the honeysuckles around my bedchamber."

Quitman's account of a typical day's events at the plantation is an agenda of amusements. "Mint-juleps in the morning are sent to our rooms, and then follows a delightful breakfast in the open veranda. We hunt, ride, fish, pay morning visits, play chess, read or lounge until dinner, which is served at 2 P.M. in great variety, and most delicately cooked in what is here called the Creole style....In two hours afterward every body— white and black—has disappeared. The whole household is asleep—the *siesta* of the Italians. The ladies retire to their apartments, and the gentlemen on sofas, settees, benches, hammocks, and often, gipsy fashion,

Dockmen load and unload steamboats at a wharf in New Orleans, the market for all the cotton and sugar produced along the Mississippi River. A sign on the steamboat Gipsy, *in the foreground, shows that it stopped at Bayou Sara, near Rosedown (Chapter Six).*

on the grass under the spreading oaks. Here, too, in fine weather, the tea-table is always set before sunset, and then, until bedtime, we stroll, sing, play whist or croquet." It was at the end of this portrait of ease that Quitman declared he had given up thinking in favor of dreaming. The plantation's atmosphere of languor quickly enveloped the visitor because all work was done out of sight, beyond the parterres and groves of oaks. It was as if the house existed in solitude and was sustained without effort.

Quitman was so charmed by plantation life that he swallowed whole the slaveholders' contention that the slaves were a happy lot. "They are strongly attached to 'old massa,'" he wrote. Visitors such as Quitman were often convinced of the happiness of the slaves because they saw only the household servants, who were better treated than the field hands, upon whom fell the hard labors of cultivation and the harsh treatment of the overseers. But as Martha Turnbull of Rosedown discovered, even the household slaves were not the contented servants they could appear to be. During the Civil War, when freedom was in the air, her household slaves went on strike: "When I ordered Celiame to scrub my kitchen she walked off and sat in her house for 3 days. Stepsy was impudent and would not cook....For 9 days Lucinda refused to come and wait on me....Augustus said he would not cut wood....Simon would not weave." She could get them back to work only when she agreed to pay them: "Julia one week at 40¢...Penny and Lancaster 2 days each, $1.60...Penny cleaning front yard, gave her 2 lbs. coffee—2 lbs. sugar."

The Civil War both destroyed plantation society and shrouded it in the romance of the permanently lost past. The Confederate dead became the ghosts of the South and part of the lore of the mansions they died to defend. Sentimental novels did much of the romanticizing, but the reality was poignant enough. In her *Memoirs of a Southern Woman* Mary Polk Branch, a relative of the Polks of Rattle and Snap, recalled the time when the body of a young general was laid out in the parlor of her family's mansion. "A bloody handkerchief was over General Cleburne's face, but one of his staff took from his pocket an embroidered one, and said: 'Cover his face with this; it was sent him from Mobile, and I think that he was engaged to the young lady.'" "No wonder," Mrs. Branch then wrote, "that it is said that the jingle of spurs and the measured tread of a Confederate soldier is often heard in the hall of the old house at night."

The war, and the economic hardships that ensued, destroyed many Southern houses, leaving them in the lonely ruin that symbolized the sufferings of the South. A ruin is always evocative, but in the South the shells of the old houses touched a deep strain of melancholy. "They are all that remains tangible of [the Southerner's] once potent civilization," J. Frazer Smith wrote in the dedication of his 1941 book *White Pillars*, a serious study of the South's architecture that is tinged throughout with nostalgia for antebellum culture. Perhaps the most famous book about Southern architecture, published in 1948, is Clarence John Laughlin's *Ghosts along the Mississippi*. In his brooding, sometimes surrealist black-and-white photographs, Laughlin set out to summon up the shades that haunted the plantations, to capture a "secret and innominate life that inhabits old houses." Nostalgia was at the heart of his undertaking, too: "And seeing these houses...we shall be taken out of ourselves; out of our own era, with its organized madness." It is ironic that the plantation should have been so thoroughly discredited and then so thoroughly romanticized. Its continuing appeal can be traced, in part, to the beauty of its architecture. It was an architecture of graciousness, practicality, pride, and sometimes grandeur. It was the architecture of great wealth suddenly acquired, "as if by magic," by men and women who then raised stately columns to demonstrate that they were the true heirs of an ancient ideal of freedom; and that made it the architecture of illusion.

1
SHADOWS-ON-THE-TECHE
TOWN HOUSE ON A BAYOU

Shadows-on-the-Teche has long been famed as one of the most beautiful houses in Louisiana. The two-and-a-half-story brick house stands amid a tranquil garden of live oaks, cedars, crepe myrtles, and camellias. The oaks tower over the house and cast the cooling shadows that gave it its name. At the end of the sloping garden behind the house, the Bayou Teche flows on its sinuous path to the Gulf of Mexico. When Shadows was built steamboats used to pass within hailing distance.

Though Shadows has a portico with Tuscan columns—eight plain shafts, painted white—and an entablature in the temple style, a Greek or Roman Revival label docs not sit easily upon this house. Shadows is a Louisiana house with some classical details, not a classical house plunked down by a bayou. Its beauty is not in any classical perfection of proportion or symmetry, but in the way the house is one with its landscape. The garden is as much a part of the ambience of the place as the house itself. It nestles up to the walls, casting its shadows and scents around and inside. By day patterns of sunlight and shade play over the house. On nights when the moon is out, the bayou glows with a silver light; canes and bamboos rustle in the breeze. About 1860 the entire house was painted white, but in the decades after the Civil War, when the house was neglected, the rain gradually washed off the paint,

Opposite: Still surrounded by the live oaks Mary Weeks planted about one hundred and fifty years ago, Shadows-on-the-Teche stands just a stone's throw from Bayou Teche in New Iberia, Louisiana. It was built as a town house by David Weeks, a sugarcane planter, in the early 1830s.

Overleaf: Shadows' lush garden was partly the work of Weeks Hall, a great-grandson of David and Mary Weeks, who lived in the house from 1922 until 1958. The gallery and loggia on the bayou side of the house were bricked up in the 1860s and reopened after a recent restoration.

revealing the warm coral bricks that had been made of soil taken from the shore of the bayou.

David Weeks built the house, in New Iberia, Louisiana, in the early 1830s as a town house, centrally located among the four scattered sugar plantations he owned.

Alfred R. Waud, a traveling correspondent for Harper's Weekly, *penciled this New Iberia street scene that appeared in the magazine in 1866. The town had been founded in 1765 by French and Spanish settlers and was originally named Nova Iberia by the latter.*

Weeks was more concerned with comfort than display. Shadows is not an imposing mansion but a cozy family home designed to catch the evening breeze from the bayou. Despite the portico on the front of the house, Shadows owes less to neoclassicism than to traditional Creole ways of building. Weeks and his builder, James Bedell, used the typical Louisiana plan of five rooms to a floor—three rooms wide and two rooms deep on the sides. There are no interior hallways; the rooms are all interconnecting, and galleries in the front and rear provide passages. The main stairway is out-of-doors in the front gallery; it runs up at one end of the colonnade and is screened by louvered panels. A narrow stairway within the house provided access to the second floor for the servants. Shadows is entirely open to its surroundings. Every room on the first two floors has a doorway to a porch.

The first story has an office, a pantry, two guest bedrooms, and, in the center, the dining room, floored with cool tiles of contrasting colors. On the second floor three bedrooms and a sitting room flank the parlor. Weeks decorated the interior plainly. He copied a parlor cornice from Minard Lafever's popular architectural stylebook, *The Young Builder's Instructor,* and installed fluted pilasters in the parlor, but in general he adorned with a very light hand.

David's wife, Mary, and their six children moved into the house, which they called Home Place, in June 1834. David was not with them because in May he had left for Connecticut for his health. Mary began adding her decorating touches to the house and planting a garden. On June 28, 1834, she wrote David describing her delight in their new home. "We have moved into the New House that I find more cool and pleasant than I had expected.... I never saw a more delightful airy house, my own room particularly. I have all the children in it and open the doors and windows every night."

On the same day David was writing to her from New Haven with disturbing news: he was thirty-five thousand dollars in debt. In her reply Mary hastened to assure him that she was not spending much money in furnishing the house and said that he should not concern himself with money but with getting well. "Our old furniture distributed about the rooms looks better than you would think. We have got from New Orleans one dozen thirteen-dollar chairs that I have put in the dining room—the chairs and bedsteads from Franklin [a nearby town] are all we have bought.... Think not that poverty has such horrors as you think. Your health is the worst we have to contend with. I would take a basket and pick cotton every day if it would do you any good." While Mary was trying to save money on furniture, David was busily buying furnishings in New Haven and sending them to New Iberia— carpets, oil lamps, mahogany and maple chairs, dressers, and washstands. His illness, now thought to have been cancer, was steadily worsening, and perhaps he felt that making these purchases was the last thing he could do to help his family. In August 1834, Weeks died in New Haven.

In spite of David's debts, Mary was able to keep Shadows and live a comfortable life with her children. The garden was her great pleasure. The live oaks that now give Shadows its distinctive character were planted by her in the 1830s. She ignored the advice of her neighbors and planted the oaks close together. With its spreading branches, this variety of oak likes a lot of room to itself. But Mary's plan worked out for the best—her oaks grew far taller than usual because they had no space in which to spread out, creating a high shady canopy. In 1833 she had written

Slaves chop and haul bundles of sugarcane in a Louisiana field as an overseer on horseback supervises their work. David Weeks owned four sugarcane plantations near New Iberia. Louisiana planters were providing about half of America's sugar by 1830.

to David's sister Rachel O'Connor, who lived on a nearby plantation, asking her to identify some flowers she wanted to plant. Rachel's reply, preserved in the Weeks family papers, gives a good picture of the way Mary's garden originally looked. "I

Overleaf: The ground-floor dining room, with its Sheraton style table, opens onto the loggia and the garden. A watercolor of the house, one of a pair completed in 1861 by Adrien Persac, is displayed to the right of the fireplace. The convex mirror over the mantel dates to the Regency period.

think the leaves that you enclosed," Rachel wrote, "are both one kind…and that the bush is *periwinkle*…. The periwinkle is a small evergreen vine that bears blue flowers, very pretty near the pickets where they can run up them." Rachel sent a hortensia with instructions on caring for it. "The flower in the box is tender and needs to set in a shady place where the sun shines but little during the summer and [must be] covered from frost in winter. The flower first appears a greenish white and in several days becomes a beautiful pink colour, which lasts some days…. I have a small cutting of the yellow rose…which I will send if it lives until next winter. Have you any of the flowering almond? I can send you some next winter if you want it." Mary spent almost every day in the garden, planting, pruning, or just sitting at the edge of the bayou with a book. She was impatient whenever the rain kept her inside— "In dark bad weather when I cannot go in the garden," she wrote in a letter, "time hangs heavy on my hands."

The mahogany double bed in the master bedroom dates from about 1845. This is one of the many antiques that were in the house when Weeks Hall died in 1958. Some were acquired by him and others may have been original family possessions.

In 1841 Mary married Judge John Moore. It may have been Judge Moore who made an alteration to the house that cut down on the ventilation David Weeks had striven for. The first-floor loggia and second-floor gallery in the rear of the house were bricked up. Perhaps the breeziness that Mary had found so delightful her first summer in the house had turned out to be intolerable in the winters.

During the Civil War Judge Moore went to Texas for safety, but Mary refused to leave. Federal troops occupied New Iberia, and General William B. Franklin took

Opposite: An upstairs sitting room, adjacent to the master bedroom, has a doorway to the rear gallery that overlooks Bayou Teche. The mahogany breakfront secretary was purchased by David Weeks on his northern trip in 1834. The sofa and card table, both in the Federal style, date from about 1815.

Overleaf: The furnishings of the second-floor parlor include an American Empire pedestal table and mahogany side chairs from the 1840s. The wallpaper has a wide border of festoons in the style of the 1840s. Invoices show that Mary Weeks bought indigo candles for the chandelier in this room.

over Shadows as his headquarters. Mary, who was in her sixties, shut herself up in the second floor of the house with her sister and daughter-in-law and disdained to have any contact with the Northerners downstairs. The army requisitioned all the food supplies in the area; the townspeople could get meat only from profiteers, who charged one hundred dollars for a ham. A friend suggested to Mary that she should take the oath of allegiance to the United States to gain favorable treatment from her unwanted guests, but she scorned the idea—"My husband and children shall never know that mortification," she said. Weakened by lack of food and the cramped conditions, she died in December 1863.

After the war the Weeks family lived at Shadows intermittently. Their sugar plantations failed with the loss of the slaves, but an immense salt dome, discovered under their fields on Weeks Island, restored the family fortune. Shadows became known as "the house that sugar built and salt saved." Weeks Hall, the great-grandson of David and Mary Weeks, was brought up at Shadows and took it over in 1922 after his return from Europe, where he had studied art. Hall rejuvenated the garden with fresh plantings of rare camellias, azaleas, boxwoods, wisterias, jasmines, honeysuckles, and thick rows of bamboos for privacy. He hired a New Orleans architect to renovate the house, but he drew the line at repainting the exterior. As an artist he realized that, stripped by the elements of its paint, Shadows had found its true beauty. It was Hall who gave the house its evocative name, Shadows-on-the-Teche.

Renowned as a host and conversationalist, Hall attracted writers, artists, and Hollywood celebrities to Shadows. D.W. Griffith filmed a movie there. The author Henry Miller visited, and wrote that Hall's great work of art was his style of life at Shadows. He said that the house was "as organically alive, sensuous, and mellow as a great tree." Before his death in 1958, Hall arranged for the National Trust for Historic Preservation to take on the restoration (they reopened the original rear gallery and loggia) and maintenance of the house. "I have lived on the place attending to it and building it," he wrote. "Nothing in life has meant, or will mean, more to me than this garden on a summer morning before sunrise."

This bedroom was used by the Weeks children. The Federal-style four-post bed, made of mahogany, dates from 1820. The oldest Weeks child, Frances, whose portrait hangs over the parlor mantel, was among the hundreds killed in 1856 when a hurricane swept over Last Island, an offshore resort.

TURNING COTTON INTO SILVER

When a New Yorker visited a Southern city in 1801, he was surprised to find that "there is no display of [silver] plate beyond the spoon and fork—and as to porcelain or chrystal [sic] services, they are totally unknown." In a few decades the cotton boom changed all that, and the table of a Southern mansion was as elegant and aglow with silver as that of any Northern magnate.

The era of Southern prosperity came as the neoclassical Empire style was giving way to the Rococo Revival, with its emphasis on elaborate relief decoration. At first most silver sold by Southern retailers was made in the East or imported from abroad. The largest Southern firm, Hyde and Goodrich in New Orleans, offered nearly every item of flatware that was available in New York. By the 1850s, cities such as New Orleans and Mobile had attracted skilled silversmiths from the East and from Europe. Among the most talented was Adolphe Himmel, a German immigrant who made the cake basket opposite and the milk pitcher on page 37 for Hyde and Goodrich. After the Civil War Southern firms had difficulty competing with companies such as Gorham in Providence, Rhode Island, which was turning out large quantities of relatively inexpensive, machine-produced wares. In the South and the North, the unmistakable gracefulness of handcraftsmanship gave way to the economy of mass production.

Made sometime between 1853 and 1861 by Adolphe Himmel, New Orleans' foremost silversmith, this Empire-style cake basket ranks among the best American examples of the form. Finely braided strands of silver make up the handle, fluting covers the surface of the basket, and a beaded border decorates the base.

This Empire-style coffee and tea service belonged to Andrew Jackson, who owned silver made by Nashville and Philadelphia silversmiths. The Rococo Revival tray was a gift from Sam Houston, who had fought under Jackson during the War of 1812 and was a frequent visitor at the Hermitage (Chapter Five).

As the nineteenth century progressed, Americans showed a fondness for increasingly specialized utensils such as the cake server above, made by James Conning. Originally from New York, Conning moved to Mobile in 1842. It was there that he made this cake server, decorating it with pierced and engraved rococo detail.

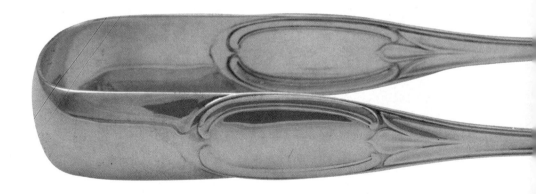

Serving tongs were another popular nineteenth-century form. The maker's marks on this elegantly simple pair indicate that it was manufactured in the Northeast and sold through Hyde and Goodrich's New Orleans store about 1850. The handles are decorated with olive-shaped cartouches.

Made by Hyde and Goodrich in the first half of the 1850s, this Rococo Revival goblet is decorated with repoussé grapes and vines. The rim of the goblet is slightly scalloped, following the naturalistic twists of the vine. Space has been left free for engraving because goblets were often presented as commemorative gifts.

Hot-milk pitchers were a popular accessory in southern Louisiana homes, where café au lait was a favorite drink. This one has an octagonal body decorated with repoussé and chased floral designs. It was made about 1852 by Adolphe Himmel and his partner, Christopf Kuchler, for Hyde and Goodrich.

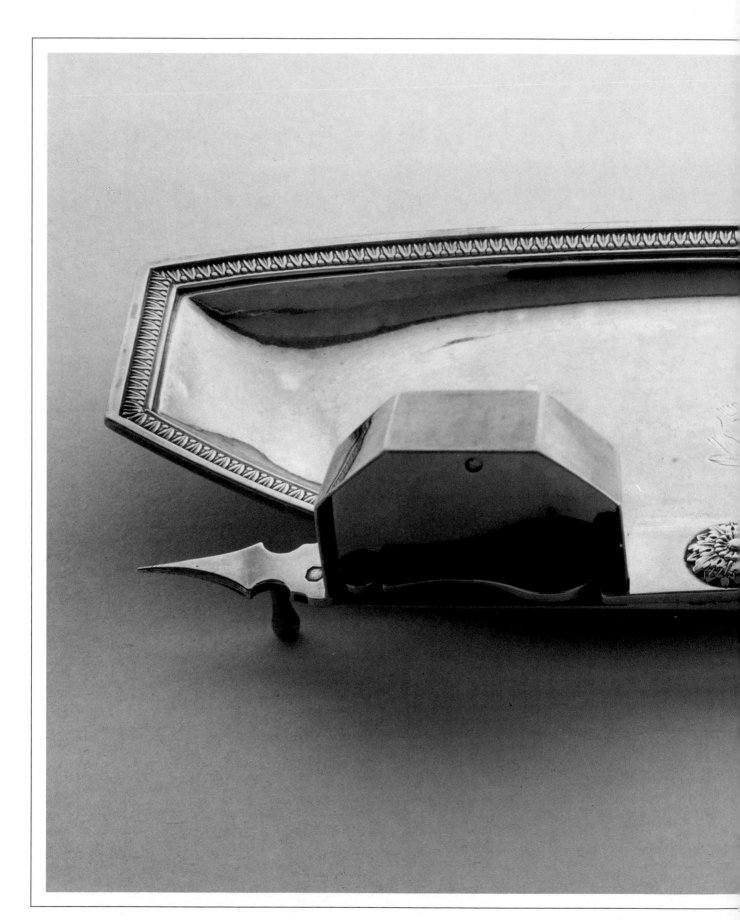

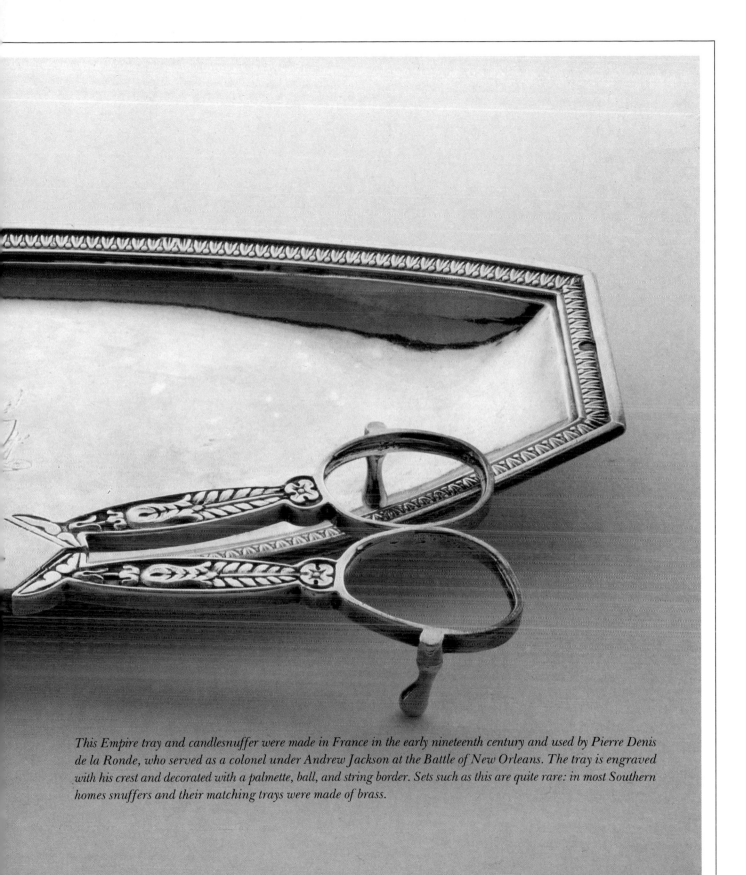

This Empire tray and candlesnuffer were made in France in the early nineteenth century and used by Pierre Denis de la Ronde, who served as a colonel under Andrew Jackson at the Battle of New Orleans. The tray is engraved with his crest and decorated with a palmette, ball, and string border. Sets such as this are quite rare: in most Southern homes snuffers and their matching trays were made of brass.

This Rococo Revival epergne, commissioned by General Nathan B. Whitfield about 1850, was the centerpiece of his dining table at Gaineswood (Chapter Two). Its removable glass dish held towering arrangements of fruit.

This detail of the epergne's pedestal shows three cherubic figures set in characteristically ornate rococo decoration. The whole piece was cast in spelter, a zinc alloy, and then silver plated.

2

GAINESWOOD

A GRECIAN SHOWPLACE

D emopolis, Alabama, was one of many Southern towns that flowered during the cotton boom in the three decades before the Civil War. The fertile black soil of central Alabama, atop a deep deposit of lime, yielded large crops and equally large fortunes. In Demopolis the planters built elegant, sumptuously furnished town houses to display their new wealth. On his plantation just outside the town, Nathan Bryan Whitfield outdid them all. From about 1843 to 1861, he worked on a Greek Revival showplace, classically simple on the outside and extravagantly decorated on the inside, with a columned drawing room for dancing, a domed parlor and dining room, and a pleasure park with a lovers' lane and an artificial lake. Perhaps his only failure was in naming his estate. "Gaineswood" simply does not evoke the romance of Whitfield's imaginative house and his dreamy, artfully contrived landscape.

Gaineswood is Greek in its decorative details but thoroughly American in its energy and freedom from stylistic stuffiness. Some nineteenth-century builders selected a specific Greek temple as a model and tried to duplicate its facade and forms faithfully. Whitfield took what he wanted from classical models and then improvised. He added new rooms and porticoes whenever money and time were available and tore apart a room when fresh inspiration struck him. The result is not

Opposite: General Nathan Bryan Whitfield spent nearly twenty years building Gaineswood,
a veritable paean to Greek architecture inside and out. For his porte cochere Whitfield used the form
of a Doric temple. The garden's small circular temple is visible between the columns.

Overleaf: The western facade of Gaineswood presents a carefully balanced composition
of colonnades. The areas enclosed by low balustrades were Whitfield's flower gardens. Near the
porte cochere he placed a marble statue of Pomona, the Roman goddess of fruit.

a confused hodgepodge but a house of delightful irregularity.

There are no fewer than four colonnades, of varying sizes and shapes, on the sides of Gaineswood. Other Southern builders, notably George Polk at Rattle and Snap (Chapter Four), used the portico as a grand display of status. Whitfield toyed with porticoes and made them a device of visual pleasure. The house is symmetrical when viewed from the west, with a four-columned portico

Just after he completed Gaineswood in 1861, Whitfield hired the artist John Sartain to engrave this view of the house and its picturesque lake. Whitfield and his second wife can be seen at bottom left, with a servant pushing a pram.

neatly balanced by flanking colonnades. The north side of the house has an entirely different character: a large portico thrusts out boldly, but a smaller, gently curving veranda next to it deflates any pretensions to classical grandeur.

Whitfield was a successful, thirty-four-year-old planter and politician in North Carolina when he emigrated with his family to Alabama's Black Earth country in 1834. Already well off, he went west in search of lands that would make him rich. Whitfield's uncle had preceded him to Alabama and in 1832 invited his nephew to come and see for himself the cotton boom that had taken hold in the state. Whitfield was impressed with the land. He wrote to his wife, Betsy, that "the lands here are *very rich . . .* and on most of them there are totally pleasant places to live." Thanks to the fertility of the black soil, cotton planters were making quick fortunes with apparent ease. "Our friends are all generally well and making money or rather getting rich rapidly. . . . Five large steamboats passing the river almost every day loaded with the rich products of the country and cotton selling at 14 cents in Mobile." A six-hundred-acre plantation worked by about sixty slaves could yield ten to twelve thousand dollars a year. Whitfield waited two years before deciding to buy—a

---

Opposite: A looking glass seems to create a second stairway in the hall of Gaineswood. The wallpaper is a reproduction of a popular mid-nineteenth-century pattern.

Overleaf: The drawing room, which Whitfield proudly called "the most splendid room in Alabama," is richly ornamented with columns and pilasters, a frieze of lush, swirling foliage, and Greek honeysuckles along the ceiling beams. A mirror set in a niche in the wall, lined up with another one across the room, reflects an infinity of chandeliers and columns.

hesitancy that cost him dearly: "The lands that Uncle James asked me $10,000 for last year, he is now asking $25,000 for."

In 1834 the Whitfields and their four children moved to a plantation near Jefferson, where they lived in a modest cabin. Within three years he owned more than four thousand acres, but the prosperity the Whitfields found in Alabama was marred by the deaths of three children and by Betsy's poor health. Whitfield did not record what illness beset his wife and carried off their children, but he began to suspect that the area was simply unhealthy. In 1843 he bought another plantation near Demopolis, fifteen miles away, from General George S. Gaines, who was the Federal government's representative to the Choctaw Indians. When Whitfield was partly through building his new house, he named it after Gaines, whose close friendship with the Choctaws had made the region safe for white settlement in the early decades of the century.

"I am in hopes that our move here has been of benefit to your mother's health," the anxious husband wrote to his daughter Mary Elizabeth—who was away at college—after the family had moved to Demopolis and settled into Gaines's simple, two-story log house. The change of scene did no good for Betsy; she would die three years later.

Whitfield began enlarging Gaines's house as soon as his family moved in. The work progressed in several stages, interrupted by his wife's bouts of ill health and by his frequent business trips to Mobile and northern cities. Like many enthusiastic house builders, he was oblivious at first to the discomforts that inevitably result from living in an unfinished house. The family apparently took them in stride as well. With good humor a daughter wrote Whitfield in Mobile to inform him of a minor disaster that occurred one winter night in 1848: "We had a very hard storm here last Saturday night....I got a complete shower bath. I woke up about midnight...in a large puddle of water." Shortly after this incident his patience grew thin, as he confessed in a letter, "I am getting quite tired of it and I hope to see it completed but this I cannot hope to see under at least 12 months more." His prediction was off by more than a decade. As the work dragged on, he could feel the years passing not only for the task but for himself. One hot summer day he penciled an inscription on

---

Whitfield used the Corinthian order, with its profuse, leafy capitals, for the drawing room's columns. The rosewood Rococo Revival furniture, which is not entirely in keeping with the classical style of the room, was made in a Northern workshop before the Civil War.

a disk of wood and placed it inside a column he was setting up in the hallway. "This cap," he wrote, referring to the column capital, "was made by Gen. Nathan B. Whitfield and put together the 9th day of August 1854 and the 54th year of his age...." The note was found during a renovation in the 1970s.

Greek honeysuckles frame an art-glass transom in the dining room. The glass, featuring vines and flowers with a Greek-key border, was painted and then baked. Every transom in the house has a different design.

Whitfield was particular about details. He based his designs for columns and plaster ornaments on authentic Greek models, using a two-volume work on Greek art, *The Antiquities of Athens* by James Stuart and Nicholas Revett, as his guide. Colorful art glass depicting mythological scenes (virtually identical glass is in the Capitol in Washington, D.C.) filled the transoms; the doorknobs were silver. The painted wallpaper, with floral patterns and parades of gold griffins, came from France. Bryan Whitfield, at medical school in Philadelphia, was given the job of obtaining marble fireplaces. He was as thoroughgoing as his father, to whom he sent the following report in 1853: "I looked at mantels at the different marble warehouses and not finding a pair of the size and price desired I engaged Mr. Struthers to make them of the best Italian veined marble at $55 for the pair." The mantels were not Bryan's only contribution to the house. In 1849, after his graduation from the University of North Carolina, Bryan had built a circular platform on the roof, surrounded by a balustrade, where the family liked to sit on warm evenings for musicales.

What Whitfield could not buy he made on the plantation. A slave named Sandy fashioned some of the decorative plasterwork. Other slaves made the columns using machinery of Whitfield's own design. A team of mules endlessly walking in circles around a shaft powered the workshops.

The dining room, one of two domed rooms in the house, has a dining table attributed to François Seignouret of New Orleans. The mahogany-veneered sideboard in the corner is in the style of Baltimore cabinetmaker John Hall. The maker of the imposing silver-plated epergne is not known.

Whitfield also laid out elaborate gardens. Slaves excavated an artificial lake in front of the house and sank a 1,100-foot-deep artesian well to feed it (another well provided running water to the house—a first in the state). He planted poplars, cedars, and crepe myrtles on two little islands in the lake, built a miniature round temple with a dome, and graded the land to create a hilly scene that would be the very essence of the picturesque. In parterres flanking the house, he planted fragrant beds of roses, flowering quinces, kiss-me-at-the-gates, bridal wreaths, and hyacinths.

This niche for toiletries is in a dressing room next to the mistress's bedroom. Gaineswood was the first house in the state to have running water, which was piped in from an artesian well.

Whitfield was determined to enjoy the house even in its incomplete state. One day in 1849 he invited five hundred guests for a party: "We had quite a crowd here yesterday," he wrote, "at a show of wild beasts and a negro sing and dinner." When his son Bryan was married a few years later, Whitfield held the reception in the drawing room, undeterred by an unfinished wall. He stretched a thirty-foot sheet of canvas over the wall and painted a row of columns. Some guests, perhaps out of politeness, later claimed to have been completely fooled by Whitfield's trompe l'oeil colonnade. Fortunately when Whitfield was married for the second time in 1857, to his cousin Bettie Whitfield, the ceremony took place in the bride's hometown, Baltimore.

When the bride and groom came home to Gaineswood, they faced another three years of painters and plasterers. At length, in 1861, Whitfield wrote with relief, "I have the house nearly complete, having just got through with the painting and papering. The parlor and dining room are also changed in effect by adding some lights [small cupolas, also called lanterns] to the ceiling, which are very beautiful. The large drawing room is now completed and I think it is the most

The Grecian theme of the house makes itself felt even in the master bedroom, where a pair of Ionic columns support a ceiling beam. The portraits over the fireplace, recently copied from nineteenth-century miniatures, show Whitfield at age twenty and his first wife, Betsy, at age eighteen.

splendid room in Alabama." To celebrate he commissioned Philadelphia artist John Sartain to engrave a portrait of the house and grounds. In the engraving, which truly captures the romantic flavor of the estate, the master and mistress can be seen strolling happily by their lake.

The completed house had twelve or thirteen rooms on the first floor (some were removed in the 1920s) and six on the second. The floor plan, like the facade, reveals that Whitfield was not a builder who was tyrannized by notions of strict symmetry. The house meanders pleasantly, with an L-shaped main hallway, three staircases, and a profusion of doorways. The master bedroom has no fewer than five doors.

One of the clever features of Whitfield's interior design is his arrangement of the parlor and dining room. They are like mirror images facing each other across a hallway. The two rooms both have domes, with identical plaster decorations of scrolls and floral motifs. The domes were both decorative and functional. At the top of the domes, Whitfield put three-foot-wide cupolas with tinted windows that cast a bluish light on the rooms.

For the parlor Whitfield designed a mechanical musical instrument in which cylinders turned by a crank directed jets of air into pipes that yielded a sound like flutes. He dubbed it the "flutina." A Brooklyn instrument maker named George Hicks built it according to Whitfield's instructions. For the dining room he ordered a spectacular silver-plated epergne and designed a rosewood cabinet with a curved, glass-enclosed compartment to store it.

The spacious drawing room, measuring twenty by thirty feet, was one of the largest rooms in Alabama. Whitfield designed it as a room for festivity. He surrounded it with twenty-eight fluted columns and pilasters topped with leafy Corinthian capitals. The coffered ceiling is lavishly adorned with plaster medallions and rosettes. From Paris he ordered vis-à-vis mirrors—a pair of mirrors facing each other across the room—that make the room appear to be a forest of columns receding into infinity. He placed statues of the goddesses Flora and Ceres in front of the mirrors, so his guests could see their multiple reflections mingling with ancient deities in the mirrors, as if they were in a vast Olympian hall. A daughter recalled that Choctaws used to gather at the windows to gape at Whitfield's parties.

The mahogany bed in the mistress's bedroom, virtually identical to the one in the master bedroom, has posts carved with pineapples and acanthus leaves. It was made in the 1830s or 1840s. One of Whitfield's daughters made the floral needlepoint picture by the bed.

The Whitfields had little time to savor the pleasures of their estate. It was finished just at the outbreak of the Civil War. They lived in fear of a sudden attack by gunboats from the Tombigbee River and in dread of the consequences should the South lose the war. Whitfield's term for the Union forces was simple and direct—the Abolitionist. "I very much fear they will destroy everything I have," he said. Though Confederate General Leonidas Polk, the Episcopal clergyman known as the "fighting bishop," temporarily made Gaineswood his headquarters in 1864, no fighting took place in Demopolis, and the house escaped damage during the war.

In 1862 Whitfield was badly injured in a fall. As he was crossing a drainage canal, a plank splintered, tossing him onto rocks in the bottom of the canal. He walked with a limp after the accident, and his health declined. He died in 1868. Before his death he had sold Gaineswood to Bryan for forty thousand dollars, which he placed in a trust for the care of Bettie and their daughter Nathalie.

The Whitfield family owned the house until the 1920s. Left empty during the 1940s, the house gradually fell to ruin, a prey to thieves, vandals, and the elements. A mulberry tree somehow took root in the dining room and grew right through the dome. Fortunately the Whitfield children and their descendants saved many of the furnishings, so when the house was named a national landmark the restorers could recreate Gaineswood's rooms with considerable authenticity. Whitfield's flutina has lost its voice to age, but modern restoration techniques may soon have it warbling waltzes and Scottish airs again. One of the Parisian mirrors in the drawing room, shattered long ago when a vandal fired a pistol at his reflection, has been replaced. The new mirror and its older twin once more reflect an infinite forest of columns, peopled by ancient gods and modern tourists.

Whitfield added this classical flourish to the mistress's bedroom—a bay with modified Corinthian columns and pilasters. The glass in the windows appears to be curved, but actually only the frames are curved. With its northeastern exposure, the bay is bright but relatively cool.

IN REGAL STYLE

In 1824 a Southern cabinetmaker advertised that his furniture "shall not be excelled by any brought from eastern cities or elsewhere." The extraordinary Mississippi-made piece opposite attests to the skill of the craftsmen working in the rural South. Plantation owners were able to obtain superb furniture from local makers to stand beside the items they purchased from fashionable shops in New Orleans, the East, and Europe. Increasing prosperity meant that houses were growing larger, and most now devoted an entire room to dining. The sideboard became the focal point of this room, just as the sofa dominated the parlor. Many sideboards, such as the one on pages 62–63, were designed along neoclassical lines, although, in keeping with an extravagance born of prosperous times, they were often too big for the room.

Most of the Southern furniture that has survived dates from the Empire and Rococo Revival periods. Empire furniture reflects a fascination with the civilizations of Greece, Rome, and Egypt. The fashion was born in France and encouraged by Napoleon, who was eager to distance himself from the stylistic trappings of the Bourbon dynasty. This period of elegant form and restrained use of ornament was followed by a return to rococo curves and rich decoration. Most Southerners kept their Empire sideboards but filled their parlors with rosewood sofas in sweeping curves made possible by new laminating techniques and machine-powered tools. As they commissioned matching sets of these pieces, richly carved with roses and vines, Southerners were recapturing the grandeur of eighteenth-century French kings.

This unusual Hepplewhite-style secretary and clothespress was made near Natchez, Mississippi, between 1795 and 1820. More than seven feet tall, it is made of walnut, with an inlay of holly or maple in a popular motif—the trailing vine. The fall-front desk provides a writing surface and storage space for papers.

Made in Georgia between 1800 and 1810, this neoclassical sideboard is fashioned of birch inset with darker walnut. Its design is unusual; the two end sections have concave curves and the center is serpentine.

The brass figure of a sphinx decorates the corner of this opulent Empire table, made in France about 1810 and now in the library at Waverley (Chapter Three). The table has a black marble top and elegant classical ornamentation. The Egyptian influence began to affect the Empire style after Napoleon campaigned in Egypt between 1798 and 1801.

Part of a set from the card room at Rosedown (Chapter Six), the graceful, open-backed chair above was made by the New York cabinetmaker John Henry Belter. He pioneered the process of laminating rosewood and bending it, with heat and pressure, into sinuous shapes, which were then decorated with hand carving.

The fire screen in the parlor at Rosedown is needlepoint enclosed in a handsomely carved rosewood frame. According to family tradition the needlepoint was done by Martha Washington. Her great-great-granddaughter married William Turnbull and brought the precious heirloom to Rosedown.

This exuberant Rococo Revival sofa was made about 1855 and has been attributed to the New York firm Joseph Meeks & Sons, which had a branch office in New Orleans. The frame is rosewood, an expensive wood imported

from Brazil or East India. The silk upholstery is a reproduction of a favorite Southern pattern. Originally purchased for a home in Nashville, the sofa is now in the music room at Rattle and Snap (Chapter Four).

This detail of another Meeks sofa at Rattle and Snap shows the lacy, pierced carving that is characteristic of his shop. Although delicate in appearance, the frame is very strong because it is formed of layers of bonded wood. The hand-carved roses and grapes are typical rococo motifs that evoke the splendor of eighteenth-century France.

3

WAVERLEY

THE OCTAGONAL VOGUE

On a bluff about a quarter of a mile from the Tombigbee River near Columbus, Mississippi, George H. Young built one of the most imaginative houses in the South—Waverley. On the outside the house does not appear to be unusual. A straight gravel path leads from a brick and wrought-iron gate to the front door, past a magnolia that is the largest in Mississippi and century-old boxwoods gnawed into odd shapes by deer. In the center of the two-story facade, made of cypress boards painted white, a recessed portico with a pair of slender Ionic columns pays nodding homage to the Greek Revival. A large octagonal cupola atop the hipped roof adds an unclassical verticality to the facade.

Waverley's entrance, decorated with red Venetian glass in the sidelights and fanlight, leads to a pleasant architectural surprise: instead of a typical entrance hall, Waverley has a broad and lofty rotunda that soars up fifty-two feet to the cupola. Only from the inside does the visitor see that the house actually has four stories, not just the three apparent on the exterior.

Three octagonal balconies surround the hall. Curving staircases ascend from the ground floor and from each balcony to the next higher, all the way to the cupola. A long rod attached to an acanthus leaf medallion in the cupola supports an ornate ormolu chandelier. The balconies and staircases, with more than seven hundred

Opposite: Colonel George Young, the owner of a fifty-thousand-acre cotton plantation near Columbus, Mississippi, built Waverley about 1850. The center of the house is occupied by an unusual octagonal rotunda, topped by the cupola. The magnolia to the left of the path is the largest in the state.

Overleaf: A pair of curving staircases ascend from the ground floor of Waverley's rotunda to the first balcony. Colonel Young designed the rotunda as a place for parties and dances, and also as an enormous chimney that would draw heat from the entire house.

mahogany balusters, are a masterpiece of woodworking. They are cantilevered, so they seem to float over the rotunda without any apparent brackets or braces to clutter their effect. The geometric precision of the octagonal balconies contrasts with the delicate arcs of the stairways to create striking arrangements of lines and curves. The geometric combinations appear in almost endless variety, changing their appearance depending on where the viewer is standing. It is a pleasure merely to stand beneath a stairway and see the way it merges neatly with the balcony overhead. This spacious, multileveled hall has an almost theatrical flavor. The rotunda and the balconies created an impressive stage set for balls and dinner parties—the light comedies of plantation life.

The rotunda has an ingenious practical effect as well: it is five to ten degrees cooler inside the house than outside, thanks to the highly efficient ventilation provided by the open central hall. The heat rises to the cupola as if through a gigantic chimney and escapes through sixteen windows. Since there are several windows in each room, every part of Waverley is always cool and breezy.

A pair of three-story wings flanks the hall. Four rooms open onto the floor of the hall—a dining room, library, parlor, and the master bedroom. There are four bedrooms on the second floor, and the third floor was used for storage. From the fourth-story cupola, which is virtually enclosed in glass, Young enjoyed a panorama of his plantation. All the rooms on the first two floors are identical in size and shape, but Young relieved the sameness with a variety of Greek, French, and other revival motifs. An Egyptian Revival bedroom on the second floor features lotus flowers on the entablatures over the windows. All the rooms on the first two floors have Italian marble fireplaces, some carved with French and Egyptian motifs. Young had his painters marbleize the baseboards and installed porcelain doorknobs and keyhole covers on many of the doors. He gave special attention to decorating the windows, ordering brass valences from France and reportedly spending five hundred dollars per window on the drapery in some rooms. A fragment of the original material has

Opposite: The dining room is furnished with an early Victorian English table and Philadelphia Chippendale chairs. Eighteenth-century Sèvres porcelain stands on the table's silver plateau. The still life over the mantel is by a contemporary Mississippi artist, Emmitt Thames.

Overleaf: Louis XV-style armchairs and sofa, with their original Aubusson upholstery, surround a marble-topped French Empire table in the library. The Chinese bowl on the table was made around 1810. The handwoven Indian carpet reproduces a Robert Adam design.

survived—a swatch of brocaded silk in peacock blue.

The four downstairs rooms have ceiling medallions, and two are decorated with plaster friezes, including a superb openwork frieze of acanthus leaves in the parlor where Young built an arched alcove for wedding ceremonies. According to an old

story, Young brought Irish artists up from Mobile to execute the plasterwork. When they discovered the extent and complexity of the project Young had in mind, they turned down the job and packed their bags. He persuaded them to stay for a meal and proceeded to entertain them so lavishly that they lingered two months as guests and then agreed to do the work.

Young was born in Georgia in 1799, attended Columbia University's law school in New York, and practiced law in Georgia until he moved his family to

The Egyptian Bedroom, named for the Egyptian-style woodwork around the windows, was used by one of Young's daughters and occasionally as a guest room. The magnificent bed (in detail on the opposite page) was made in France in the eighteenth century.

Mississippi in 1835. He served as secretary to the government commission that was auctioning lands that had belonged to the Chickasaw Indians. Young purchased some land himself, gave up practicing law, and devoted himself to cotton planting—with great success. Eventually he owned about fifty thousand acres, and his plantation included orchards, sawmills and gristmills, warehouses, and a store. He ran a fishing operation and a ferry service on the Tombigbee River. Before his wife died in 1852, just as Waverley was being finished, she bore ten children.

In his book *White Pillars*, published in 1941, J. Frazer Smith refers to a diary Young kept. According to Smith, the diary says that Young hired an Italian architect named Pond to design Waverley. Smith's research turned up records of a St. Louis

Opposite: The rosewood bed in the Egyptian Bedroom was entirely carved by hand. The four posts are spiraled, and the pierced headboard and footboard are decorated with acanthus leaves. The bed hangings are reproductions of nineteenth-century designs.

Overleaf: Another bedroom at Waverley has a set of Rococo Revival furniture made by Elijah Galusha of Troy, New York, who had once worked for John Henry Belter. The set, fashioned of rosewood about 1850, includes the bed, the marble-topped dresser between the windows, and an armoire.

architect, Charles I. Pond, who may have been the man, despite the fact that his name is not Italian. Unfortunately Young's diary has disappeared, and it is impossible to verify Smith's account.

The octagonal shape of the cupola and the balconies in the central hall was not an individual bit of whimsy. When Young built Waverley, in the early 1850s, octagonal houses were a fad. The vogue for octagons had been touched off by Orson Fowler, who published a book entitled *A Home for All* in 1848 in which he propounded the superiority of the octagonal form over the rectangular. The book was an immediate hit: in just

A second-story balcony at the rear of the house offers a view of Waverley's garden, framed by Ionic columns. The boxwoods directly below the balcony were planted by Colonel Young when he built the house in the 1850s.

eight years it went through nine printings. The octagon, Fowler said, is close to the sphere, which is the perfect shape devised by nature itself. Not only is the octagon a more practical shape, it is more beautiful by far than the square or rectangle. "Why continue to build in the same SQUARE form of all past ages?" Fowler asked rhetorically. "Nature's forms are mostly SPHERICAL.... Then why not apply her forms to houses?... The octagon, by approximating to the circle, incloses more space for its wall than the square, besides being more compact." Sharp angles, "sticking out in various directions," he wrote, offend the eye, whereas the rounded shape is far more agreeable. Fowler's messianic prose style (which is also wielded in books promoting phrenology, vegetarianism, and sex education) brought forth a wave of octagonal houses in the North. Southern octagonal houses are rare, though octagonal features such as Waverley's balconies and cupola are more common.

Most likely Young and his architect were influenced by Fowler's book; but they played a variation on the octagonal theme. Waverley is a conventional rectangle on the outside—Young put the octagon inside the house. Most octagonal houses centered on a small spiral staircase lit by a cupola. Young realized the possibilities of using the octagon to create amusing visual effects within a soaring interior space worthy of a church.

Waverley was a splendid setting for a party. A local doctor who wrote a history of

Columbus in the nineteenth century described Young as a man whose "genial manners and unstinted hospitality drew a constant stream of visitors." Another local historian wrote that Waverley was a "many-roomed house equal to the entertainment of [Young's] sons and daughters and their families, and his numerous friends." He kept a kennel of hunting dogs on the plantation and built a brick swimming pool, fed by an artesian well, in the garden behind the house.

Young died in 1880, and an unmarried son, William, took over the house. In 1893 William gathered a group of hunting enthusiasts in the library and founded the National Fox Hunting Association. After William's death in 1913, Waverley stood empty until the 1960s. Though nearly all the furniture was scattered in the intervening decades, the house was spared modernization. The original room arrangements, decorative plaster, painted graining on the doors, and magnificent stairways and balconies were all preserved virtually intact. Another fragile decorative element that survived is the red Venetian glass, with lyre-shaped muntins, surrounding the front door.

The current owners of Waverley, who bought the house and began to restore it in 1962, have furnished it with eighteenth- and nineteenth-century antiques—the sort of furniture the Youngs might have owned. In the master bedroom, for example, there is a half-tester bed by Prudent Mallard, the New Orleans furniture maker whose pieces can be found in many plantation mansions, and a Rococo Revival sofa-and-chair set by John Henry Belter. The library is furnished with a Louis XV sofa and four armchairs with Aubusson coverings. A few of the original furnishings remain: three mirrors with elaborate gilt frames, four French ormolu chandeliers, and a walnut secretary in the library. A mirror in the rotunda is cracked, but the restorers chose not to replace it because it was broken in Young's time. At a party someone placed a lamp too close to it and the heat cracked the glass. The hall chandelier has six glass globes, once lit by gas made on the plantation by heating pine knots in a retort. Leaves decorate the whole chandelier, and from the foliage six ormolu faces—handsome young men—look down on the floor, like shy lads gaping at a splendid dance they are too timid to join.

Overleaf: With its octagonal balconies, Waverley's rotunda is one of the most spectacular halls in the South. The multiple levels of octagons, surrounded by delicate balustrades, contrast with the curves of the staircases to produce a delightful visual effect.

THE SOUTH
IN MINIATURE

Southerners were often accused of spoiling their children. In the early nineteenth century an Italian visitor to America wrote, with evident disapproval, that "the fathers, especially in the South, yield sadly and foolishly to their children . . . whose wishes they do not restrain." Before the Civil War the fond parent who wanted to indulge a child would most likely have purchased costly imported toys: as late as 1903 there were fewer than a hundred toy makers in all of the United States. One of the largest toy makers in the world was the W. S. Reed Company of Leominster, Massachusetts, which made the steamboat on pages 92–93.

Rocking horses such as the one opposite are by now the quintessential symbol of childhood. Riverboats, not surprisingly, were extremely popular with Southern children. With their paddle wheels, smokestacks, and great walking beams, steamboats were probably more exciting to play with than today's toy racing cars and dump trucks. Baby and doll carriages were also modeled after adult vehicles. The two shown on pages 96–97 reveal something of the character of the Victorian era—they faced the baby forward to impress the public. (In the 1920s designers turned the body of the carriage around so that the baby faced its mother.) Pared down to simple forms and colors, toys such as these have all the evocative power of folk art. They are fresh and amusing icons of their age.

With a realistic mane and a long-suffering look, this rocking horse from the nursery at Rosedown (Chapter Six) may have been ridden by several generations of Turnbull and Bowman children. Until recently, when it was carefully restored and repainted, the horse bore many signs of loving use.

These demurely dressed dolls from the nursery at Rosedown were made in the late nineteenth century and belonged to one of the Turnbull grandchildren. The heads, hands, and feet are made of composition—a mixture of papier-mâché and wood pulp—that has been painted by hand. The doll on the left was recently restored, but the one on the right is in its original condition.

Made of lithographed paper and wood, this side-wheeler was manufactured about 1880 by the W. S. Reed Company. Diners are having a meal in the grand saloon above the paddle wheel, while other passengers peer out of cabin windows. Between the two smokestacks is the walking beam, which on a real boat linked the engine to the crank shaft that turned the paddle wheels.

Minstrel figures balance precariously on the bow and stern of "Reed's Latest Sensation," manufactured in 1881. The men dance and play their instruments as the toy is pulled along. The lithographed paper covering the boat includes vignettes of a plantation house and a levee scene.

The three-wheeled baby carriage at left, from the nursery at Rosedown, was made in England about 1840. Designed to resemble the horse-drawn carriages that transported adults, it bears a family crest on the side and a collapsible hood. The handsomely painted doll carriage above, with its fringed canvas sunshade, was made about 1885 and belonged to Mary Elizabeth Whitfield, one of General Whitfield's granddaughters at Gaineswood (Chapter Two).

MAMMA'S CONSERVATORY.

CHILDREN PLAYING WITH FAWNS.

PEOPLE FROM ALL CO

A SCENE IN AUSTRALIA.

THE CHURCH STYLE.

HARRY WITH HIS BIRDS.

THE INDIAN ADJUTANT.

AN OMNIBUS.

THE CHINCHILLA.

A BOAR HUNT.

THE OSPREY, OR SEA EAGLE.

A HORSE AT PLAY IN A MEADOW.

THE SOLITARY HERON.

OH DEAR! WHAT SHALL I WRITE?

THE WREN.

GREAT EXHIBITION, 1851.

THE NIGHTINGALE.

A MAN ATTACKED, BY WOLVES.

THREE PRETTY DUCKS.

MEDITATING HIS WORK ON "ANIMATED NATURE."

MARY'S FIRST LESSON.

D

Scenes of animals and birds predominate in this tattered but colorful fragment from a nineteenth-century children's book at Tulip Grove, a Jackson family mansion near the Hermitage (Chapter Five). Although its title page and cover are missing, the book appears to have been published not long after the Crystal Palace exhibition in London in 1851, which drew people from all over the world to admire the fruits of the Industrial Revolution. A view of the exhibition appears on the right page of the book.

4

RATTLE AND SNAP

A PRIDE OF COLUMNS

Twenty-three-year-old George Washington Polk came to Tennessee from North Carolina in 1840 not as a doughty pioneer but in the grand manner of a young lord, to take possession of land his father had given him. A work force of slaves preceded him to make the way straight, building a comfortable, nine-room frame house for Polk and his new bride, Sallie Hilliard. Not long after Polk arrived, he began work on a magnificent mansion, Rattle and Snap. That unusual name was probably derived from the way in which Polk's father obtained the land. According to family tradition, Colonel William Polk won about 5,600 acres from the governor of North Carolina in a game of chance called rattle 'n snap, which was more than likely a dice game. Colonel Polk was a distinguished veteran of the Revolutionary War—he served with Washington at Valley Forge—and a close friend of Andrew Jackson's. He was a leader in politics and in the early development of Tennessee. With his good political connections, he prospered in Tennessee land deals and eventually owned about 600,000 acres.

Four of Colonel Polk's sons—George, Leonidas, Rufus, and Lucius—received portions of the Rattle and Snap tract and emigrated from North Carolina in the 1830s and 1840s. Three of them built mansions on their properties. (Rufus died soon after settling in Tennessee.) The three great houses—Rattle and Snap, Hamilton Place, and Ashwood Hall, which burned down in the 1970s—stood close to one another where the brothers' properties came together, near Columbia. That town was the home of the most eminent member of the family: cousin James K. Polk,

George Washington Polk, a wealthy Tennessee cotton planter who was a cousin of President James K. Polk, built Rattle and Snap in the 1840s. One of the finest mansions in the South, Rattle and Snap was recently restored and completely refurnished with antiques of the 1840s and 1850s.

the eleventh president of the United States, whose house was quite modest compared to those of his wealthy cousins in the country.

In the first half of the nineteenth century, many superb mansions were built in middle Tennessee, but none surpassed George Polk's Rattle and Snap; indeed it remains one of the most magnificent houses anywhere in the South. A monumental colonnade stretches across the front, with ten fluted columns of carved wood, four of them set forward under a pediment so that the house seems to advance on the visitor. It was an imposing display of status. As the Tennessee architectural historian John Kiser remarked, "Most people of the time had four columns, Andrew Jackson managed to have six, but George Polk had ten." Acanthus-leaf Corinthian capitals top the columns, but the colonnade is more an outburst of romantic fervor than an allusion to classical antiquity. This is one of the Southern mansions that depart from the strict norms of classical proportion and use

A cast-iron porch and balcony on the eastern wall add a Louisiana flavor to this Tennessee mansion. The metalwork—with rococo vines, Greek honeysuckles, and slender Gothic columns— is as fine as that on a New Orleans town house.

columns as a symbol of Southern wealth and pride. True Greek Revival buildings adhere to the classical ideal of achieving beauty through symmetry and proportion, but Rattle and Snap harks back to the baroque passion for drama and the illusion of movement on a massive scale. The colors of the building add to its vitality. The white columns stand out against a sandy facade, while the side walls are unpainted brick. The pale blue ceiling of the portico creates the illusion that the columns soar into the open sky. Polk added an oddly delicate feature to the eastern wall of the house—a

Opposite: Polk installed his stairway to the side of the entrance hall, thus making the hall spacious enough for entertaining. The Gothic Revival clock was acquired from Hamilton Place, the plantation owned by George Polk's brother Lucius. The clock's case may have been made at that plantation.

Overleaf: With its gilt mirrors, arrangements of Rococo Revival furniture, and dramatic window drapery of lace, taffeta, and damask, the double parlor expresses the exuberant decorating style of the South at mid-century. The carpet was specially made from English patterns of the 1830s.

beautiful cast-iron balcony with slender columns, rosettes, and twisting, rococo vines.

Polk used columns inside the house with an equally grand effect. There are two Corinthian columns in the entrance hall. The stairway, an important element in many Southern mansions and one that was often used as an architectural flourish, is tucked away in a side hall. Two more Corinthian columns divide the double parlor, which is fourteen feet high and runs the length of the house. These columns bring the monumentality of the facade into the house, which was conceived not as a cozy rural retreat but as a palatial series of spaces for entertaining lavishly. Not only did he build a huge double parlor, he also insisted on a large dining room, where twenty-four guests could sit down to dinner.

The house is shaped like a reversed L. The main section contains the double parlor, dining room, entrance hall, and a music room for family gatherings. Three bedrooms and a library are upstairs. A wing at the back provided workrooms and quarters for the house servants on the first floor, and guest bedrooms on the second. An uncommon amenity at Rattle and Snap was its bathroom, with running water and a lead-lined wooden bathtub with a drainpipe leading to a tiled underground waste tunnel.

Though Polk's original furnishings were dispersed before the end of the nineteenth century, the interiors seen today at Rattle and Snap have been restored with great care to re-create the mansion's atmosphere of elegance. The walls of the entrance hall and double parlor are painted pale gray and beige to set off ornate gilt mirrors and the rich colors—deep reds and golds—of the carpets, drapery, and Rococo Revival furniture.

If Polk hired an architect to design the house, no record of him exists. Polk may have acted as his own architect. As Jill Garrett, the Maury County historian, points out in her unpublished history of Rattle and Snap, George Polk was a well-educated man who had studied at the University of North Carolina and owned a library of

Opposite: A sliding door has been opened to combine two dining rooms into a banquet room. The dining table came from an Alabama plantation, and the chairs were made by François Seignouret of New Orleans. The silver-and-glass chandeliers are reproductions based on an 1840s pattern. The sideboard is in the style of New York furniture maker Joseph Meeks.

Overleaf: The Rococo Revival sofa and chairs in the music room originally belonged to a Nashville family. This set, the Aubusson rug, and the Steinway-Webber rosewood piano date from about 1850.

two hundred books, an impressive number for rural Tennessee in the 1840s. George's brother Leonidas, who designed Ashwood Hall himself, probably advised George on the construction and may have been responsible for the elements that

An informal sitting room in the upstairs hall is furnished with six chairs by Joseph Meeks in the same pattern ordered by President Polk for the White House. The sofa is by Antoine Gabriel Quervelle, who also made furniture for the White House.

show a Louisiana influence. The cast-iron balcony on the side of the house recalls those of New Orleans and early-nineteenth-century Savannah; the wide, short entrance hall with the stairs in a side corridor is another Louisiana trait. Because he was the Episcopal bishop of Louisiana, Leonidas spent a lot of time in the Deep South and knew its architecture well.

In a letter one of Polk's daughters wrote that skilled workers among Polk's slaves built Rattle and Snap with stone and bricks they made on the property. The carved wooden columns and cast-iron capitals, made in Pittsburgh, Cincinnati, and Louisville, were shipped by river to Nashville and hauled to the house by ox-drawn wagons. The rear wing of the house was finished by the fall of 1843, because family records report that the Polks' second child was born there in October of that year. By 1845 the entire mansion was complete.

Polk styled himself a grandee—his house alone is evidence of that—and he was the host of some of Tennessee's most splendid dinners, balls, and horse races. None but the wealthiest of men could live on such an opulent scale; Polk kept his head above water by borrowing. Despite the generally prosperous times, the cattle, corn, and cotton he raised on his plantation could not get him out of debt. Even before the Civil War and the emancipation of the slaves, he was in deep financial trouble.

The entire Polk family were staunch Confederates. Leonidas, a general in the Confederate army, was killed in 1864 in Georgia when a cannonball struck his chest.

A second-floor bedroom has a half-tester bed in the style of the New Orleans cabinetmaker Prudent Mallard, with curved outlines, restrained rococo ornamentation, and a boldly carved cartouche on the headboard. The popular half-tester, or half-canopied, beds were also known as "Arabian beds."

Two of Polk's sons were Confederate officers. George Polk did not serve during the war—he suffered from rheumatism—but like many noncombatants he was quite bellicose. In 1861 he wrote to his son James, "I should like to have a turn at those rascally Lincolnites by the side of my boys. Save me a horse and a sword if you can. I will try and get a pair of pistols in Nashville." A year later Mrs. Polk wrote to James, "Your Pa feels like getting up a company of 100 men...but I do *fear* for him to do so because his health is so *feeble*."

Rattle and Snap is rich in Civil War lore. Federal troops marched into the area in April 1862, with at least some of them intent on pillaging the houses of the "nabobs," as the Northerners referred to the plantation masters; but the Southerners had taken precautions. As Polk's niece later wrote, "The war had been going on but a short while when people began talking about what to do when the Yankees came, what to hide and where to hide it." Polk's youngest daughter, Caroline, recalled that her family had hid the silver within one of the huge columns by lowering her brother, who held a basket with the silver, inside the column on a rope. According to one story, also recounted by Caroline, the Yankees galloped up to Rattle and Snap with thoughts of torching it. But in the hall an officer spotted a portrait of George wearing his Masonic ring. The officer reported this to his superior, a Mason himself, who ordered the house placed under guard.

Polk's fortunes collapsed in the 1870s. He was forced to sell Rattle and Snap to pay his debts and suffered the ignominy of living for nearly two decades within sight of his old home. He died in 1892. Like many Southern mansions Rattle and Snap had its lean years. At one point it was a hay barn; there are photographs of the house that show hay bulging from the windows. Private owners saved the house from total ruin in the 1950s. The current owner, who bought the house in 1979, embarked on an ambitious restoration effort that included archaeological digs on the grounds, paint and wallpaper studies, and the acquisition of a major collection of Victorian furnishings. Once a shabby hulk on its hill by the highway, Rattle and Snap once again proclaims George Polk's vision of Southern grandeur. Polk had to live out his days tantalized by that colonnaded vision, perhaps pondering the risks of an enterprise that had begun with a roll of the dice.

In George Polk's time a thick grove of trees stood between the mansion and the road; today there is a clear view of fields and distant hills. During the Civil War the opposing armies skirmished along this road—a Confederate soldier was killed at the foot of the drive, but Rattle and Snap escaped damage.

PERSAC'S LOUISIANA

As the South's plantation society was reaching its full flower, a French artist named Marie Adrien Persac was traveling the Mississippi River taking commissions to paint some of the region's great houses. About twenty of his fragile watercolors have survived, including an interior of a Mississippi River steamboat (pages 116–117) and a view of a plantation store (pages 120–121), and they offer a rare and evocative picture of the antebellum years. Persac was born near Lyons, France, in 1823 and, according to family tradition, was drawn to America by a desire to hunt buffalo. Like many newcomers struggling to establish themselves, Persac is known to have tried a number of careers, listing himself at various times in New Orleans directories as photographer, architect, artist, and civil engineer. In 1869 Persac started his own "Academy of Drawing and Painting" on Camp Street in New Orleans, where he taught students to paint landscapes and portraits in oils; but no oil paintings by him have survived, nor have any buildings been attributed to his designs. His first and best-known work was a map published in 1858 showing the plantations along the lower Mississippi from Natchez to New Orleans. As he traveled to collect information for the map, Persac began painting the Louisiana views shown here, combining spacious, precisely rendered backgrounds with an almost playful use of figures cut from magazines and pasted to the surface of his paintings. These figures are handsomely costumed and so artfully chosen and arranged that the overall effect is something like an elaborate stage set, ready at any moment to come to life, dogs barking, horses stepping high.

These two views, done in gouache (opaque watercolors), are the only known Persac works showing the front and back of a house. In the top view pasted figures representing friends and neighbors animate a carefully rendered portrait of the street facade. The bottom view shows the house as seen from the bayou, tucked away behind a curving picket fence and partially hidden by the lush landscape.

Painted in 1861, this is the earliest known view showing the interior of a Mississippi River steamboat, and it is extraordinary in its detail, from the bracketed Gothic ceiling to the painted floorcloth. Persac gave this watercolor to his wife on their tenth wedding anniversary. According to family tradition, this is the boat on which they took their wedding trip, and the figures at left represent the artist and his bride.

A peacock roosts in a tree at left in Persac's painting of the Hope Estate, near Baton Rouge. The house, complete with octagonal

pigeonniers *in the French taste, was built by Daniel Hickey and his son about 1795.*

John Dominique's general store was located near Donaldsonville, Louisiana, some thirty miles south of Baton Rouge. It was supplied by Mississippi River steamboats and probably served sharecroppers from a number of nearby plantations, since major landowners would have had their own stores. In Persac's charming scene the dog in the foreground seems about to give chase to ducks in the pond.

Persac's painting of Magnolia Plantation shows smoke rising from the lumber mill, and cows and sheep, scattered over the

landscape. The house was built for the Martin family, who bought the land in 1845.

5

THE HERMITAGE

OLD HICKORY'S RETREAT

ndrew Jackson does not enjoy the reputation of a tastemaker; he carefully nurtured his image as "Old Hickory," the warrior and log-cabin democrat, the backwoods man of the people. But as his supporters were treading the halls of the White House in their muddy boots, Jackson was setting the precedent for high-style architecture in the West. His home mansion, the Hermitage, was one of the earliest Greek Revival houses built in Tennessee. As the home of the most famous man in America, the Hermitage itself became well known through engravings in newspapers—its dramatic colonnade impressed the planters of the South and the West and helped popularize the Greek Revival style.

The Hermitage was built in three stages. Jackson erected the original structure, a Federal-style brick building, between 1819 and 1821. In 1831 two wings and a one-story portico were completed; and after a fire in 1834 Jackson rebuilt the house with a new colonnade. The Hermitage as it is seen today dates to this reconstruction of 1835–1836. The large-scale, templelike colonnade is not really an integral part of the structure; viewed from the side it looks like the addition that it is. In fact the Hermitage is as much a Federal building as a

Ralph E. W. Earl painted this portrait of Jackson in 1830. In the distant background Earl painted the original brick Hermitage, with its white portico.

Andrew Jackson's Hermitage features a Greek Revival colonnade added in 1836. The original mansion, completed in 1821, had a simple Federal-style facade. In the Greek Revival style, rectangular fanlights replaced arched designs of the Federal period because the ancient Greeks did not use arches.

Greek Revival one. This dual spirit is also apparent inside: the Greek Revival facade leads to a traditional, eighteenth-century floor plan in which the stair hall bisects the house. To the left are a double parlor and the dining room wing; to the right, bedrooms and Jackson's library wing.

The landscape of ancient Greece surrounds the hallway in the scenic wallpaper made by Joseph Dufour of Paris. The mahogany sofa, upholstered in horsehair, was where Jackson usually took his after-dinner nap.

The decoration of the hall continues the classical theme of the facade. The walls are covered with French paper, made in Paris by Joseph Dufour, showing scenes from the Greek tale of Telemachus' search for his father, Odysseus, after the Trojan War. The paper depicts an episode—not a part of Homer's epic *The Odyssey* but a later composition—in which Telemachus journeys to the island of the enchantress Calypso in his search for his father. Full of lush semitropical beauty, expanses of blue ocean—an exotic sight for Tennesseans—and colonnaded palaces, the brightly colored paper entirely surrounds the large room and even continues up the wall along the staircase. The architectural historian Kenneth Severens has suggested that the story of Telemachus—a son's quest to find his heroic father—appealed deeply to Jackson because he never knew his own father, who died a few weeks before Jackson was born. He ordered this paper not once, but three times.

The double parlor was furnished in 1835 by Jackson with help from his adopted son, Andrew Jackson, Jr., and young Andrew's wife, Sarah. They bought Empire furniture, some of which is still in the parlor today alongside Rococo Revival furniture Andrew, Jr. bought in 1851. One item of great sentimental value to Jackson, in the rear parlor, is a mahogany center table presented to him in New Orleans after his victory over the British there in 1815. The grateful citizens had given a party for Jackson and his wife, Rachel, in a room furnished with pieces by the city's finest craftsmen. When the party was over, they presented the Jacksons with the entire suite of furniture.

For his dining room furniture, Jackson went to the New Orleans shop of François Seignouret. A page from Jackson's account book shows that in May 1821, when he was en route to Florida to take up his post as provisional governor, he stopped in New Orleans and purchased a variety of furnishings from Seignouret for almost a thousand dollars. Among the purchases listed in the account are the dining table and sideboard in the Hermitage dining room. The sideboard is an impressive piece, in the monumental scale of the Empire style, with two massive pillars, a large star in a central panel, and three claw feet descending from clusters of leaves. In contrast to the stylishness of the Seignouret furniture is the dining room's carved hickory mantel—a tribute to Old Hickory from one of the veterans of the Battle of New Orleans. The old soldier worked on the mantel only one day each year, the anniversary of the battle—January 8. It took him twenty-four years to finish the piece, which Jackson installed in the dining room on January 8, 1840. The silver entrée dishes in the dining room once belonged to the naval hero Stephen Decatur. After Decatur was killed in a duel, Jackson bought some of his silver from Decatur's widow, who was in financial difficulties.

Judging from a remark in one of his letters, Jackson was not an admirer of elaborate furniture. "I like plain solid wood better than the carved wood, as it can be more easily kept clean," he wrote to Sarah with disarming practicality. His library reflects this taste for the simple. The bookcases, made of cherry, have no decorative carvings, nor does his desk. When he read his many newspapers in the morning, he liked to sit in a plain but comfortable campeachy chair—a type of chair he would have seen in New Orleans. Though Jackson preferred the plain to the elaborate, and was a lifelong advocate of economy, he did not stint when it came to furnishing his household. He surrounded himself with fine things—New Orleans furniture, French wallpaper, costly silver.

Nashville was a nine-year-old cotton town on the Cumberland River when Jackson settled there in 1788. Tennessee was still a part of North Carolina, and Jackson arrived, at the age of twenty-one, as the state's public prosecutor for the district. He established a successful private law practice and prospered in land speculation, cotton planting, and breeding racehorses—his finest horse, Truxton, won twenty thousand dollars in purses. He was elected Tennessee's first congressman in 1796, and the following year he won a special election for the Senate. He

Overleaf: Rococo Revival chairs, purchased by Andrew Jackson, Jr. in the 1850s, furnish the double parlor along with Empire-style items from President Jackson's time. The circular table in the rear parlor was presented to the elder Jackson by the citizens of New Orleans after the War of 1812.

resigned from the Senate and then accepted an appointment as a circuit judge. In the courtroom he won respect as an evenhanded judge who charged his juries to put fairness above concern for the letter of the law.

In 1791 he married Rachel Donelson Robards. The circumstances of the marriage would dog him throughout his political career. Rachel was separated from her first husband when she and Jackson decided to marry, and she believed that her divorce had become final before her second marriage took place; but actually it had not. Technically Rachel was guilty of bigamy. The Jacksons had a second ceremony performed after the divorce became final, but the supposedly scandalous marriage was fuel for Jackson's political enemies, who posed as the defenders of family values and labeled Rachel "a profligate woman" and "a wench" in newspaper articles and speeches.

François Seignouret made this Empire-style sideboard in the dining room. It is often said that Seignouret fought in Andrew Jackson's army at the Battle of New Orleans, but no evidence has ever been found to substantiate the story.

In the 1790s the Jacksons lived on two plantations in the Nashville area, first at Poplar Grove and then in a mansion at Hunter's Hill. Financial reverses forced him to sell Hunter's Hill in 1804 and move into a log house on a farm of 420 acres, which he called the Hermitage. While the Jacksons were living in the log house, they adopted a nephew of Rachel's, naming him Andrew Jackson, Jr. They would have no children of their own.

In 1819 Jackson began building the first brick Hermitage, a substantial, rectangular house in the Federal style, with two stories and eight rooms. A visitor from Charleston, a city of elegant homes, found much to admire at the Hermitage: "You enter a large and spacious hall or vestibule, the walls covered with a very splendid French paper, beautiful scenery, figures, etc.... To the right are two large and handsome rooms furnished in fashionable and genteel style as drawing rooms—

The mantel in the dining room was carved by a man who had served under "Old Hickory" during the War of 1812. Appropriately enough, he made his tribute to the general out of hickory. It was installed in 1840. Jackson purchased his dining table from Seignouret in 1821.

rich hangings, carpets, etc. To the left is the dining room and their chamber. There was no splendor to dazzle the eye but everything elegant and neat."

It was in these "fashionable and genteel" chambers that Jackson and his political advisers, the so-called Nashville Junto, devised the campaign image of Jackson as

the backwoods democrat. The War of 1812 had made Jackson a national figure—he was lionized for his stunning, seemingly impossible victory over the British at the Battle of New Orleans. His popularity won him the majority of the votes in the 1824 election, but he did not gain a majority in the electoral college and lost the vote in the House of Representatives. In 1828, after a vigorous campaign, he won the office with ease. As he was preparing to leave

The log house above was one of several Jackson built when he moved to the Hermitage property in 1805. This one served as a guest house. Jackson and his wife, Rachel, lived in a log house until 1821, when the brick mansion was completed.

for the inauguration in Washington, Rachel died suddenly. To the devoted husband it was a terrible loss, and he brooded upon it for years.

In 1831 Jackson decided to enlarge the Hermitage and hired the architect David Morrison, a Pennsylvanian who had come to Nashville in 1828 to build the state penitentiary, to undertake the project. Morrison designed other important public buildings in Nashville as well, including the Union Bank and the Methodist Church. His work—symmetrical, restrained, even a bit severe—reflected a Georgian spirit more at home in the eighteenth century than in the nineteenth.

To Jackson's Federal-style house Morrison added two wings for a dining room and library and adorned the front with a fine Grecian portico. Morrison chose the plain and dignified Doric order for the colonnade, which was one story high with a two-story, pedimented section in the center marking the entrance. On the frieze over the columns, he placed laurel wreaths, a popular motif that symbolized military triumph; on this house the symbolism was appropriate. Morrison's nicest touch was in the garden, where he built a monument over Rachel's tomb—a small-scale Greek temple, circular in shape, and topped by a copper dome.

Jackson was in Washington while the renovation was going on, so a friend, General John Coffee, stopped by the house occasionally to see how the work was going. A letter he wrote to Jackson in April 1831 about the project shows how architecture was yet a relatively simple science in which amateurs still felt they could dabble. "Your mechanics were at work on the improvements…on the mansion house. I took the liberty of suggesting some immaterial alterations in the addition, which was approved by the projector of the building, who said he would consult you about it." General Coffee felt no reluctance to alter the building plans in the midst of construction. As it turned out his suggestion was a good one and was adopted by Jackson and Morrison. He proposed extending the library wing by twenty feet to add a room for the plantation's overseer and a covered walkway to connect that room with Jackson's library. The enlargement of the house made it possible for Jackson to invite Andrew, Jr. and his bride, Sarah York of Philadelphia, to take up residence at the Hermitage.

The Greek Revival style was already in the air, but Morrison's Hermitage retained some of the schematic formalism of past decades. Though his portico was a gracious adornment, the whole facade was a careful composition of separate elements—wings, colonnade, portico, elevated pediment—that followed Palladian principles. An accident, ironically, erased Morrison's well-ordered design and cleared the way for something new. In October 1834, just after Jackson had left for Washington after vacationing at the Hermitage, the house caught fire. The flames started in a chimney, sparks ignited the roof, and, because no one could find a ladder in time, the fire spread through the building. Andrew, Jr. and the field hands were able to save some of the furniture, and Jackson's papers, from the first floor. With the exception of the dining room

An 1831 map of Nashville included this engraving of the Hermitage, which had been redesigned that year by the architect David Morrison. His colonnade burned down in 1834. To the right is the Grecian temple built by Morrison over Rachel Jackson's grave.

wing, the interior of the house was extensively damaged; but the brick walls remained sound.

Jackson was philosophical when he heard the news. "The Lord's will be done,"

Overleaf: Jackson died in this bedroom on June 8, 1845, at the age of seventy-eight. Only nine days before, George P. A. Healy had finished the portrait of Jackson by the bed. The room is furnished as it was when Jackson died, with his personal articles still on the dresser and night table.

he wrote to his son. "It was He that gave me the means to build it and He has the right to destroy it, and blessed be his name. Tell Sarah to cease to mourn its loss. I will have it rebuilt." The citizens of New Orleans, with refreshed memories of Jackson's defense of the city during the War of 1812, proposed a national subscription to pay for the reconstruction of the Hermitage. Jackson declined the offer, and asked that any funds already raised be given to charity.

Andrew, Jr. oversaw the reconstruction of the house, with the help of his father's friends Colonel Robert Armstrong and Colonel Charles Love. They hired the Nashville builders Joseph Reiff and William Hume. At someone's suggestion—the records do not show whose—Reiff and Hume put a more imposing colonnade on the front. They used fewer columns, reducing the number from ten to six, but increased their height from one story to two. Instead of Morrison's plain Doric order, Reiff and Hume chose the richer Corinthian style, with acanthus-leaf capitals made of carved poplar, and fluted shafts. The colonnade presented the builders with a problem—somehow they had to accommodate the old arrangement of windows in the front wall. In the end they decided they could not space the columns according to proper classical regularity because the shafts would have blocked the views from some of the windows. As the reconstruction was going on, another of Jackson's amateur superintendents, Major William Lewis, proposed the excellent idea of increasing the height of all the rooms, in keeping with the latest fashion. This final version of the Hermitage, with its spacious rooms and soaring Greek columns, had a wide influence—the fact that the president had built a Greek Revival house gave the style instant popularity in the West.

Andrew, Jr. and Sarah assisted in obtaining new furnishings to replace those lost in the fire, including another copy of the Telemachus wallpaper for the hall. Telemachus had an additional adventure before he finally arrived at the Hermitage. Sarah ordered the paper from a Philadelphia importer, for $120, and it was shipped by steamer to Nashville. The steamer caught fire at the Nashville docks. The ever-

Opposite: The library has one of the most important historical items in the house—the marble-topped table on which Jackson wrote his commands before the Battle of New Orleans. The cherry bookcases hold some of his five hundred books. Jackson was an avid reader of newspapers, which were mailed to the Hermitage from all over the country. "His study is loaded with piles of them," said one visitor.

Overleaf: The bedroom of Jackson's granddaughter Rachel Jackson Lawrence has its original wallpaper and walnut furniture. The portrait over the mantel shows Rachel at age eighty in 1913.

vigilant Colonel Armstrong made inquiries about the shipment and was appalled to find that the paper, undamaged by the fire, had been put on the block at a salvage auction. He traced the paper to a Mr. W. G. M. Campbell and paid a call to the gentleman's home. If Armstrong's account of his visit is accurate, Campbell hastily put the paper up when he found out that someone else had a claim to it. "When I called on Campbell," Armstrong wrote to Jackson, in some heat, "I expected to get the paper; that night he cut it and put it on the walls.... My dear sir, when you have this whole matter explained it will give you a pain to find men so lost to all honorable feelings as to retain that which does not belong to them. It is a theft." The shippers agreed to purchase another set of the same wallpaper, so Telemachus had his homecoming after all.

Jackson came home to the Hermitage in 1837, after his second term. His retirement was not an easy one. According to the biographer Marquis James, Jackson left the White House with ninety dollars in his pocket and gave some of that to a friend who was broke. Andrew, Jr. had not been a good manager of the plantation. He was free with money, and Jackson felt bound to honor his son's debts, even though he was not legally required to do so. One day a store clerk appeared at the Hermitage with a bill of more than three thousand dollars. After running his eye over the account and saying, "I see many things here he could have done without," Jackson immediately wrote out a check, on borrowed funds. With his unfailing good manners, he invited the bill collector to dinner and gave him a tour of the house.

Jackson died at the Hermitage in June 1845 and was buried beside Rachel in the garden temple Morrison had built. Like two other presidents—Washington and Jefferson—who built their own houses and were unhappy away from them, Jackson had always felt uncomfortable in the capital and wished for nothing more than to be at home. Once, when he was in Washington, he wrote, "How often [do] my thoughts lead me back to the Hermitage. There in private life, surrounded by a few friends, would be a paradise."

All the items in the Hermitage kitchen, which is behind the main house, are original. The drum-shaped object by the window is a butter churn. Like many plantation masters, Jackson entertained on a large scale—at one outdoor dinner his kitchen provided for several hundred guests.

GARDENS
IN PERFECT ORDER

I n all the world," wrote one horticultural historian, "gardening has seldom found so auspicious an environment as on the cotton plantations of the South." To the uncommonly rich soil and long growing season provided by nature, Southern gardeners brought a discerning eye for landscape design. The gardens at Rosedown Plantation (Chapter Six) and Afton Villa in St. Francisville, Louisiana, are among the loveliest in the South, and both are the work of women. Rosedown's gardens (pages 143–147) are the creation of Martha Turnbull, who, like many Southern planters, followed the seventeenth-century French style based on wide avenues of trees, clipped boxwood hedges, and formal parterres. She had visited Versailles on her honeymoon and decided to bring the same elegance to her plantation setting. Mrs. Turnbull tended and improved her garden for almost sixty years, keeping a detailed journal in which mundane entries recording weather and plantings are enlivened with phrases such as "Garden looks deplorable," or, occasionally, "my Garden in perfect order." At nearby Afton Villa's garden (pages 148–153), Susan Barrow's travels and reading inclined her toward the new taste for romantic, naturalistic design. She visited Gothic Revival estates along the Hudson River and studied Andrew Jackson Downing's new and popular work, *The Theory and Practice of Landscape Gardening*. Her grounds included a formal parterre by the side of the house (page 148), but she also devoted space to a park, in keeping with Downing's view of what the ideal garden should express—"the spirit of nature . . . softened and refined by art."

Rosedown Plantation. *The gardens at Rosedown extend right up to the house, with azaleas, clipped boxwood hedges, and neat brick paths. The fountain in the background—with the figure of a cherub riding on a seashell— is one of many sculptures that the Turnbulls imported from Europe for their garden.*

In a secluded clearing at the center of the south garden, Martha Turnbull placed a summerhouse constructed of fine latticework

that creates shady privacy. Three varieties of azaleas surround the house.

A gravel path, planted with a wide-leafed variety of liriope, azaleas, and live oak trees, meanders west to the edge of the property. The original gardens covered five acres on either side of the house. Eventually twenty-eight acres were landscaped, threaded through with curving walkways.

Southeast of the house is a reservoir planted with cypress trees and azaleas. When restoration of the gardens began, many plants and shrubs described in Martha Turnbull's garden book were no longer available commercially and had to be propagated from specimens surviving on the property.

Afton Villa. *In the formal parterre, four beds bordered in boxwood have been planted with white pansies. On the left is an unusual brick-red azalea that numbers in the thousands on the plantation and has come to be called Afton Villa Red. This one has been trimmed to blend with the formality of the garden.*

To the right of the pansy beds is a boxwood maze interspersed with tulips. This formal part of the garden is the first of a series of terraces that sweep down to a ravine. The other levels, originally planted with annuals and perennials, are now grass, creating a more formal look that reveals the garden's careful design.

South of the parterre is a less formal part of Afton Villa's garden, which Susan Barrow intended to be much like a nineteenth-century English park. Giant live oak trees dominate the landscape. Today, as in the nineteenth century, visitors wandering in the park will find peacocks roaming the grounds.

Afton Villa's old cedar trees, which date from Susan Barrow's time, have been planted with wisteria vines that wind their way through the branches and bloom among the Spanish moss. There are several varieties of azaleas, including the resplendent Afton Villa Red, sometimes also called the "Pride of Afton."

Off the main drive leading to the house is an old roadbed lined with live oaks, azaleas, and gardenias. Afton Villa's half-mile

allée is the longest in Louisiana and unusual because of its serpentine design.

6
ROSEDOWN
THE PAST
INTACT

rom the road Rosedown offers only a glimpse of itself, a flash of white at the end of a long, shadowy alley of live oaks. The facade that comes into view as you walk beneath the oaks is a carefully balanced and modest composition, one that does not rise to grand heights of display like Rattle and Snap nor present a whimsical face like Gaineswood. Though it stands a short distance from the Mississippi River, in St. Francisville, Louisiana, Rosedown bears a certain resemblance to the Georgian mansions of the East—sober and reserved on the outside, and elegant within. It was built by Daniel Turnbull, one of the wealthiest cotton planters in the South, and his wife, Martha, an imaginative and indefatigable gardener.

The mansion consists of a rectangular main section, completed in 1835, and two small wings, added in 1844. Double galleries with six plain, unfluted Doric columns on each level extend across the entire front of the main section. A ground-floor gallery, open to the evening breezes, was a common feature of Louisiana houses from the time of the early French settlements. Substantial houses such as Rosedown provided the extra comfort of an airy, private, second-story gallery. Turnbull seems to have been of two minds about the design of Rosedown. The central portion displays some late-eighteenth-century decorative details—arched fanlights over the two front doors and reeded woodwork around the windows—but the classical entablature and Doric columns of the double gallery show the influence of neo-classicism. The contract for the construction of the house, with the local builder

A magnificent alley of live oaks leads to the entrance of Rosedown, the St. Francisville, Louisiana, mansion built by Daniel and Martha Turnbull in 1835. The facade combines galleries—a traditional Louisiana feature—with typical Georgian elements, such as the arched fanlights.

Wendell Wright, is a puzzling document. It describes "a frame house...to be complete after the most modern stile [and] the front to be executed with full gretio dorric collums." The phrase "full collums" suggests that Turnbull originally wanted columns of full height; and "the most modern stile" of the period, the "gretio," or Grecian, cited in the contract, certainly called for full-height columns. After he signed the contract Turnbull must have had second thoughts about a colossal Greek order.

When he added the wings in 1844, Turnbull yielded to the prevailing fashion and made them Greek, in a more accurate Doric order than the front. The wing on the southern side of the house was a particular success; it echoes the proportions of a Greek temple even though it lacks a pediment. To knit together the whole composition, Turnbull extended the balustrade of the front galleries around the eaves of the wings. A curious change in style between 1835 and 1844 is subtly evident here: the balusters on the main part of the house are heavily proportioned, almost squat, whereas the balusters on the wings are more slender and delicate, paralleling a change that had occurred in the late eighteenth century from the Georgian mode to the early neoclassicism of the Adam style.

Perhaps the most interesting feature of the mansion is its furnishings. Because Rosedown remained in the hands of the Turnbulls' direct descendants until the 1950s, nearly all of the furnishings are original to the house. Thus Rosedown gives a remarkably authentic picture of the way a wealthy nineteenth-century family deco-

The wallpaper in the entrance hall, made in 1828 by the French artist Joseph Dufour, depicts exploits of Charlemagne's knights. It is a replacement of the original paper, which was also by Dufour.

Opposite: A doorway with reeded woodwork and with delicate tracery in its fanlight and sidelights leads to the entrance hall. The curving staircase, made of Santo Domingo mahogany, seems to disappear into the ceiling. The dining room lies just beyond the hall.

Overleaf: The dining room furnishings, including the portrait of Martha Turnbull by Thomas Sully, are almost all original to the house. The overhead fan is a punkah, a type introduced to the South by English settlers who had seen it in India. It became known in the South as the "shoofly."

rated a Southern mansion. The furnishings present a mixture of styles—Empire, Regency, Rococo Revival, and Gothic Revival. In their dining room the Turnbulls had a Regency table with Duncan Phyfe chairs and a marble-topped French Empire serving table. For the parlor they purchased a set of chairs and a sofa from Prudent Mallard of New Orleans. (Daniel Turnbull's account books for 1855 and 1856 show that he paid Mallard about $3,700 for furniture, fabric, rugs, wallpaper, a clock, curtains, and tablecloths.) The parlor set is in Mallard's restrained version of the rococo style—the carving on the back of the sofa is not as elaborate as the work of John Henry Belter, and the armchairs in the set have hardly any carving at all. The Turnbulls gave pride of place to a fire screen with a needlepoint scene reputedly worked by Martha Washington—they placed it in an ornate frame by the parlor fireplace. The needlepoint came to the family through the Turnbulls' daughter-in-law, Caroline Swanick Butler Turnbull, a great-great-granddaughter of Martha Washington.

The most impressive furniture at Rosedown is the Gothic Revival bedroom suite on the first floor, which Turnbull purchased from a Philadelphia cabinet-maker, Crawford Riddell. The set had been commissioned by friends of Henry Clay's who hoped to see it installed in the White House after the election of 1844; but when Clay lost the election, the furniture was left in a warehouse. Shortly after the election Turnbull bought the set—a bed, armoire, six chairs, bureau, cheval mirror, octagonal table, and two washstands—for $1,280. Riddell was one of the finest furniture makers in Philadelphia, and the "Henry Clay" suite is among the most spectacular examples of the Gothic Revival style ever made. The monumental canopied bed, fourteen feet high, is a virtual cathedral of sleep, richly adorned with spires, small Gothic arches, and castellations on the top. Next to this bedroom Turnbull installed a bathroom with a shower; when he pulled on a chain, an overhead tank released a cascade of warm water.

Daniel Turnbull was the son of a Scottish immigrant, John Turnbull, who

Opposite: In the parlor, with its sofa and chairs by Prudent Mallard, the Turnbulls had a peerless conversation piece—the needlepoint fire screen on the left, stitched by Martha Washington.

Overleaf: The music room was the center of Rosedown's hospitality—here the Bowman sisters entertained callers with harp and piano recitals. The piano was made by Jonas Chickering and purchased by the Turnbulls in 1841. The portrait over the piano is of Eliza Pirrie, James Bowman's mother; John James Audubon painted it in 1821 when he was the Pirrie family tutor.

settled in Mississippi and made his fortune trading with the Choctaw Indians. In 1783 he moved to what is now West Feliciana Parish, where Daniel was born about 1800. Martha Turnbull was a Barrow, perhaps the most prominent name in the region. Some of the most splendid sights in West Feliciana were the Barrow mansions: Afton Villa, Ellerslie, and Greenwood. One of Martha's brothers built a seventy-five-foot steamboat, the *Nimrod*, solely for fishing and hunting trips. The boat included a stable for a dozen horses, and kennels for Barrow's hounds.

Daniel and Martha's wedding, in November 1828, united two great fortunes. For their honeymoon they took the Grand Tour of Europe. Upon their return they began to buy land for Rosedown Plantation. The name, according

This swan-neck cradle, beautifully carved out of mahogany, is an English piece dating from 1820. It was purchased during the restoration of Rosedown and stands in an upstairs bedroom.

to family tradition, was inspired by a play they had seen in New York, set at a mansion called Rosedown. In the fall of 1834, Turnbull hired Wendell Wright to build the mansion. A note in Turnbull's diary for November 3 signaled the start of construction: "Commenced hauling timber cypress for house." Wright worked quickly, and the house was finished the following May.

The tract of land where the Turnbulls built Rosedown had been occupied since the late 1700s. The impressive alley of oaks leading to the house was probably planted by the Mills family, who lived on the site until 1808. They selected sturdy young trees from the surrounding forest and transplanted them along a carriage drive. With the Mills' oak alley as her starting point, Martha laid out twenty-eight

Opposite: Prudent Mallard made the rosewood half-tester bed in the master bedroom. The Turnbulls purchased it as part of a bedroom set that included a double armoire, bureau, and commode.

Overleaf: The Turnbull and Bowman children (there were thirteen in all) played in this nursery on the second floor. The toys—a rocking horse and a doll—and the furniture are original to Rosedown. The child's desk is made of walnut, as is the canopied bed. The doll perches in a Sheraton-style high chair made in England in the early nineteenth century.

acres of gardens. Close to the house she planted parterres modeled on the gardens at Versailles, which she and Daniel had visited on their honeymoon. Away from the house the parterres give way to a luxuriant park with winding paths threading through camellias, azaleas, catalpas, crepe myrtles, and sweet olives.

The Turnbulls had three children. One son, James, died of a fever in 1843; their other son, William, drowned in a boating accident in 1856. Their daughter, Sarah, was the belle of the parish—her suitors were innumerable, and proposals of marriage numbered an even one hundred, as the story goes in St. Francisville. In the late 1850s Sarah finally said yes to James Pirrie Bowman.

Daniel Turnbull died in 1861, but Martha could look to her daughter and her son-in-law for support during the Civil War. Many of their slaves ran off to join the Union forces; the ones who remained refused to work until they were paid—a great shock to Martha. After the war she was one of the many Southern planters who adopted the sharecropping system; her former slaves rented land from her and paid her with a portion of their crops. Life was not as easy as it had been before the war, but Martha, Sarah,

The massive Henry Clay bed, over thirteen feet high, was too large for any of Rosedown's rooms, so Turnbull built a new wing to accommodate it. The bed has finely carved Gothic motifs.

and James carried on the antebellum customs of genteel society as best they could.

Sarah and James had ten children in the 1860s and 1870s, eight of them daughters. Sarah called the girls "my daisy chain" and drilled them in all the polite feminine virtues. When visitors called she expected the girls to enter the parlor one by one with a bow. Only after the first girl had finished greeting each guest could the second enter, and so on until all eight had presented themselves properly. "Miss

---

Rosedown's Gothic Revival bedroom suite was originally intended as a victory gift for Henry Clay after the 1844 presidential election. When Clay was defeated Daniel Turnbull purchased the set from its maker, Crawford Riddell of Philadelphia.

Sarah was provoked if a single girl arrived out of turn," one guest remembered. (Southern manners must have required extraordinary patience of the guests as well as the hosts.) There were enough daughters for a fine chamber ensemble—each played a musical instrument and sang. As the girls grew into women, these polite gatherings for music and chat evolved from a display of proper training into the ritual of courtship. Young men arrived in groups to clutch teacups and submit to familial inspection. One suitor won the hand of a Bowman girl who died not long after the wedding; he returned to Rosedown's parlor and won a second Bowman bride. One other daughter married, but the other five remained single.

Though their training had been in the gentler accomplishments, the Bowman daughters revealed their mettle when the time came. Martha Turnbull died in 1896, and Sarah Bowman in 1914. The boll weevil wiped out any hope of making a living from cotton on West Feliciana's already depleted soil, but in the wake of the destructive little weevil came tourists willing to pay a modest charge to see a genuine old plantation and Martha Turnbull's famed gardens. James Bowman, a distinguished, white-haired artifact of the earlier age, lived until the 1920s and graciously performed the duties of guide whenever visitors made their way up the carriage drive past the AUTOMOBILES FORBIDDEN sign. Wearing overalls and wielding shovels, the four surviving Bowman sisters (Corrie, Isabel, Sarah, and Nina) tended the garden, the vegetable patch, the chicken coop—and kept the bankers at bay. The writer Harnett Kane visited Rosedown in the 1940s and found the Bowmans content with their lot, still living "a life of a bygone period." For every portrait the Bowmans told a life story; every stick of furniture had a past. Their conversation, Kane noted, "is compounded of polite agreement and amiable expressions.... Miss Sarah taught them all the graces; hardly one has been forgotten."

The last sister died in 1955, and her heirs sold the plantation to a Texas family who restored the mansion and the grounds and opened them to the public as one of the most splendid examples of antebellum Southern life. Though the new owners had to undertake extensive repairs of the house, much of their work had, in a sense, already been done for them by the Bowman sisters, who carried Rosedown intact into the twentieth century and never knew they were living in a museum.

Viewed from Rosedown's front porch, the carriage drive and its canopy of live oaks, which are almost two centuries old, extend to a Greek Revival gateway. Martha Turnbull landscaped the grounds with a neat parterre around the house and a twenty-eight-acre park beyond.

Acknowledgments

The Editors would like to thank the following people for their assistance: H. Parrott Bacot, director, Anglo-American Art Museum, Louisiana State University, Baton Rouge; Sheila Biggs, Gaineswood; Ormond Butler, manager, Rosedown; Mrs. Fletch Coke, research chairman, The Ladies' Hermitage Association; Bill Cullison, Southeastern Architectural Archives, Tulane University; Mrs. Steven Dart, St. Francisville, Louisiana; De La Salle Provincialate, Christian Brothers of the South and Southwest; Amon Carter Evans, Rattle and Snap; Stanton Frazar, director, Historic New Orleans Collection; Sally Gant, Museum of Early Southern Decorative Arts, Winston-Salem; Jill K. Garrett, Maury County historian; Anne Hamilton, Rosedown; Patricia L. Kahle, assistant director, Shadows-on-the-Teche, a museum property of the National Trust for Historic Preservation; John W. Kiser, University of Tennessee, Nashville; Arthur Lemann, Jr., Palo Alto Plantation, Donaldsonville, Louisiana; Patricia McWhorter, assistant curator, Historic New Orleans Collection; Richard Marvin, assistant curator, Historic New Orleans Collection; Shereen Minvielle, director, Shadows-on-the-Teche, a museum property of the National Trust for Historic Preservation; Steven Rogers, Tennessee Historical Commission, Nashville; F. William Satterwhite, Nashville; Mrs. Betty Sheehan, supervisor of interpretation, The Ladies' Hermitage Association; Roberta Smith, Gaineswood; Robert and Donna Snow, Waverley; Jackson R. Stell, historic house and museum coordinator, Alabama Historical Commission, Montgomery; Edith Thornton, supervisor of administration, The Ladies' Hermitage Association; Mrs. Genevieve Trimble, New Orleans; Jan White, head of photography, Historic New Orleans Collection; Ola Mae Word, public relations director, Rosedown.

Credits

Page 8: Museum of the City of New York. 9–10: Jay P. Altmayer Collection, photo by Larry Cantrell; from *The Civil War/Brother Against Brother,* © 1983 Time-Life Books, Inc. 11: Collection of J. B. Speed Art Museum, Louisville, Kentucky. 12: Chicago Historical Society. 13: Helga Photo Studio. 14: Tulane University Art Collection. 20: The Historic New Orleans Collection. 21: Glenbow Museum, Calgary, Canada. 31: Prather Warren/Anglo-American Art Museum, Louisiana State University, Baton Rouge. 32–33: Paul Rocheleau/Ladies' Hermitage Association. 34–37: Prather Warren/Anglo-American Art Museum, Louisiana State University, Baton Rouge. 38–39: The Historic New Orleans Collection. 40–41, 47: Paul Rocheleau/Gaineswood. 61: Prather Warren/private collection. 62–63: Collection of the Museum of Early Southern Decorative Arts, Winston-Salem, North Carolina. 64–65: Paul Rocheleau/Waverley. 66–67: Paul Rocheleau/Rosedown. 68–71: Paul Rocheleau/Rattle and Snap. 89–91: Paul Rocheleau/Rosedown. 92–93: Bill Holland/Collection of Bill Holland. 94–95: Bill Holland/Barenholtz Collection. 96: Paul Rocheleau/Rosedown. 97: Paul Rocheleau/Gaineswood. 98–99: Paul Rocheleau/Ladies' Hermitage Association. 115: Collection of Shadows-on-the-Teche, New Iberia, Louisiana, a museum property of the National Trust for Historic Preservation. 116–119: Anglo-American Art Museum, Louisiana State University, Baton Rouge. 120–121: Louisiana State Museum. 122–123: Jan White/Brothers of the Christian Schools of Lafayette, Louisiana. 125: Ladies' Hermitage Association. 132: Donna Lucey Wiencek. 133: Tennessee State Library and Archives.

All other photos, including the cover, are by Paul Rocheleau.

Index

Page numbers set in *italic* refer to captions.